Studies in Computational Intelligence

Volume 1097

The series "Studies in Computational Intelligence" (SCI) publishes new developments and advances in the various areas of computational intelligence—quickly and with a high quality. The intent is to cover the theory, applications, and design methods of computational intelligence, as embedded in the fields of engineering, computer science, physics and life sciences, as well as the methodologies behind them. The series contains monographs, lecture notes and edited volumes in computational intelligence spanning the areas of neural networks, connectionist systems, genetic algorithms, evolutionary computation, artificial intelligence, cellular automata, self-organizing systems, soft computing, fuzzy systems, and hybrid intelligent systems. Of particular value to both the contributors and the readership are the short publication timeframe and the world-wide distribution, which enable both wide and rapid dissemination of research output.

Indexed by SCOPUS, DBLP, WTI Frankfurt eG, zbMATH, SCImago.

All books published in the series are submitted for consideration in Web of Science.

Nguyen Hoang Phuong · Vladik Kreinovich
Editors

Deep Learning and Other Soft Computing Techniques

Biomedical and Related Applications

Springer

Editors
Nguyen Hoang Phuong
Artificial Intelligence Division
Thang Long University
Hanoi, Vietnam

Vladik Kreinovich
Department of Computer Science
University of Texas at El Paso
El Paso, TX, USA

ISSN 1860-949X ISSN 1860-9503 (electronic)
Studies in Computational Intelligence
ISBN 978-3-031-29446-4 ISBN 978-3-031-29447-1 (eBook)
https://doi.org/10.1007/978-3-031-29447-1

This Springer imprint is published by the registered company Springer Nature Switzerland AG
The registered company address is: Gewerbestrasse 11, 6330 Cham, Switzerland

Preface

Maintaining health is an important part of human activity, in which a lot of progress is being made all the time. Artificial Intelligence (AI) and Computational Intelligence (CI) techniques—in particular, deep learning techniques—have contributed to many of these successes. However, there are still many important and crucial medical challenges. To solve these remaining challenges, we must go beyond the existing techniques. In particular, we must go beyond the currently used AI and CI techniques: we must modify the existing techniques, combine them with other techniques, and/or come up with completely new ideas. This biomedical-inspired search for new AI and CI techniques is the main focus of this book.

In line with this focus, most of the chapters describe how AI and CI techniques can help in solving medical challenges. These chapters form Part one of this book. Of course, many challenging problems remain. To solve such problems, it is important to continue developing new AI and CI techniques. If these techniques are successful in other challenging application areas, there is hope that these techniques will be helpful in biomedical applications as well. In accordance with this reasoning, in the following parts of the book, we describe AI and CI techniques that have been successful in other application areas: finance (second part), transportation engineering (third part), and physics, in particular, physics of computation (fourth part). New promising AI and CI ideas that have not yet led to successful practical applications are described in the fifth part.

We hope that this book will help practitioners and researchers to learn more about computational intelligence techniques and their biomedical applications—and to further develop this important research direction.

We want to thank all the authors for their contributions and all anonymous referees for their thorough analysis and helpful comments.

The publication of this volume was partly supported by Thang Long University, Hanoi, Vietnam. Our thanks to the leadership and staff of this institution for providing crucial support. Our special thanks to Prof. Hung T. Nguyen for his valuable advice and constant support.

We would also like to thank Prof. Janusz Kacprzyk (Series Editor) and Dr. Thomas Ditzinger (Senior Editor, Engineering/Applied Sciences) for their support and cooperation with this publication.

Hanoi, Vietnam
El Paso, USA
December 2022

Nguyen Hoang Phuong
Vladik Kreinovich

Contents

Biomedical and Other Human-Related Applications of Soft Computing

Type-2 Fuzzy Relations: An Approach towards Representing Uncertainty in Associative Medical Relationships

Bassam Haddad and Klaus-Peter Adlassnig

Abstract The acquisition of precise values such as symptoms, signs, laboratory test results, and diseases/diagnoses for expressing meaningful associative relationships between medical entities has always been regarded as a critical part of developing medical knowledge-based systems. After the introduction of fuzzy sets, researchers became aware of the fact that a central problem in the use of fuzzy sets is constructing the membership function values. The complication arises from the uncertainty associated with assigning an exact membership grade for each element within the considered fuzzy set. Type-2 fuzzy set handles this problem by allocating a different fuzzy set to each element. This paper addresses the subject of medical knowledge acquisition and representation by proposing consistent interval type-2 fuzzy relations in the context of fuzzy inclusion as a measure of representing the degrees of association between medical entities. The concept of interval type-2 fuzzy relation will be introduced to represent the uncertainty and vagueness between medical entities.

1 Introduction and Motivation

Associative relationships between *medical entities* such as symptoms, signs, laboratory test results, and diseases/diagnoses can be established in different ways. Medical knowledge has been formally represented by several *symbolic and/or numerical,* or *data- and knowledge-driven* methods, all of which have been used successfully to a certain extent (Fig. 4). An associative relationship between a symptom s and a disease d might be expressed in *two types* of measures: the *necessity of occurrence* of

B. Haddad
Department of Data Science and Artificial Intelligence, University of Petra, Amman 11196, Jordan
e-mail: haddad@uop.edu.jo

K.-P. Adlassnig (✉)
Institute of Artificial Intelligence, Center for Medical Data Science, Medical University of Vienna, Spitalgasse 23, 1090 Vienna, Austria
e-mail: klaus-peter.adlassnig@meduniwien.ac.at; kpa@medexter.com

Medexter Healthcare GmbH, Borschkegasse 7/5, 1090 Vienna, Austria

N. H. Phuong and V. Kreinovich (eds.), *Deep Learning and Other Soft Computing Techniques*, Studies in Computational Intelligence 1097,
https://doi.org/10.1007/978-3-031-29447-1_1

s with d, and the *sufficiency of occurrence* of s for d. In our context, the necessity of occurrence[1] may be interpreted as *backward implication* from d to s, i.e., $(s \leftarrow d)$, while the sufficiency of occurrence[2] may be viewed as *forward implication* from s to d, i.e., $(s \rightarrow d)$. Similarly, negative associative relationships may also be specified as backward implication or forward implication, $(s \leftarrow \neg d)$, and $(s \rightarrow \neg d)$.

The above relationships may be extended by considering multi-valued implications $\in [0, 1]$. For example,[3] the *necessity of occurrence* may be represented as multi-valued backward implication $(s \overset{\mu}{\leftarrow} d)$, *with* $\mu \in [0, 1]$, and the *sufficiency of occurrence* as forward implication $(s \underset{\mu}{\rightarrow} d)$, *with* $\mu \in [0, 1]$.

Human *expert knowledge* can be used to obtain values expressing *possible degrees* of uncertainty, while *statistical data* may add to the respective body of knowledge. These aspects have been successfully employed in representing medical relationships between symptoms, signs, laboratory test results and diseases in the differential diagnosis support systems CADIAG-I [1–4] and CADIAG-II [5–11]. In CADIAG-II, fuzzy set theory and fuzzy logic were used to represent the inherent unsharpness of linguistic medical terms by fuzzy sets, and to represent partial truths of medical relationships between these terms. Here, the *frequency of occurrence* and the *strength of confirmation* correspond to the *necessity of occurrence* and the *sufficiency of occurrence*. In addition, *negative associative implications* were considered to specify a *strength of exclusion*. *Semi-automatic statistical analyses* were thus able to support the knowledge acquisition process [12, 13].

However, creating a *solid medical knowledge base* is not always a straightforward process. It may be fraught with various problems, such as:

- Uncertainty: Exact values for expressing meaningful associative relationships are not easily obtained. Human expert knowledge and statistical data analyses might support this process. However, even quantitative medical information is never 100% accurate. Fuzzy systems have, in fact, superseded conventional methods in a variety of scientific applications. However, type-1 fuzzy systems, whose membership functions are type-1 fuzzy sets, are able to cope with uncertainties. Type-2

---

[1] Necessity: The occurrence of symptom s is said to be necessary for a disease d, then the occurrence of d guarantees the occurrence of s; e.g., s: *"increased serum glucose level"* is obligatory for d: *"diabetes"*:

$$(s \leftarrow d)$$

[2] Sufficiency: The occurrence of symptom s is said to be sufficient for the disease d, then the existence of s guarantees the occurrence of d; e.g., s: *"the detection of intracellular urate crystals (tophi)"* confirms d: *"gout by definition"*:

$$(s \rightarrow d)$$

[3] For example: The occurrence of s: *"increased serum glucose level"* is necessary for d: *"diabetes"*, however s confirms *"diabetes"* only with 0.65:

$$(s \underset{1}{\longleftarrow} d, s \underset{0.65}{\longrightarrow} d)$$

fuzzy systems relying on type-2 fuzzy sets try to handle these uncertainties further by assigning fuzzy sets or any interval $\subset [0, 1]$, defining possibilities for the primary membership. Furthermore, representing human expert knowledge and interpreting statistical data about ignorance (what is unknown?) should also be considered. *Experts' agreements* or *disagreements and conflicts among the various sets of alternatives* concerning the relevance of data [14] play a certain role in increasing the degree of uncertainty.
- Inconsistency: Medical knowledge bases containing a large quantity of relationships among medical entities might be affected by inconsistencies and incompleteness. The quality of knowledge must be ensured by appropriate checking.

The aim of the present paper is to present the basic principles of dealing with the above-mentioned aspects of uncertainties. We focus on the following aspects:

- Employing interval type-2 fuzzy relations to handle vagueness and uncertainty.
- Employing an interval-type-2-fuzzy-relation-based inclusion measure to represent binary associative relationships between medical entities. This inclusion measure corresponds to uncertain and imprecise implication relations; i.e. binary fuzzy rules. The direction of inclusion measure corresponds to the *necessity* and *sufficiency of occurrence*, which can be represented as uncertain backward and forward implication relationships. The intervals express the *possible degrees* of inclusion of one fuzzy set in another related fuzzy set, e.g., $(s \xrightarrow[I(s,d)]{} d)$, and $(s \xleftarrow[I(d,s)]{} d)$, where $I(s, d)$, and $I(d, s) \subseteq [0, 1]$. $I(s, d)$ and $I(d, s)$ represent the *uncertainty* about *a fuzzy rule*. Here, the concept of interval type-2 fuzzy relation was adopted to reduce computational complexity.
- The knowledge base of such rules should be consistent. In other words, only relationships with consistent uncertainties expressed in the form of consistent intervals are considered.

1.1 Related Work

Interval-valued techniques have been suggested by many researchers for representing uncertainty and incompleteness. Zadeh [15] proposed type-2 fuzzy sets, whose membership functions themselves are specified by fuzzy sets. This step was necessary to consider the possible uncertainty of fuzzy set functions themselves [16–19]. Baldwin [20] proposed the assignment of necessity and possibility support boundaries to *logic programs* in order to consider *uncertainty*. Turksen [21] employed compositional operations in connection with conjunctive and disjunctive normal forms to handle approximate reasoning.

The concept of *fuzzy inclusion* has been addressed by some researchers [22–24]. It has been employed in some areas of computing, such as *image processing* and *natural language processing* [25]. Helgason and Jobe [26] focused on perception-based reasoning, utilizing medical quantities such as *necessary causal ground* and *sufficient causal ground* extracted from fuzzy cardinality. In addition, the different enhancements of the CADIAG-II medical fuzzy decision support system [6, 10, 13] are closely connected to the proposed model in the sense of considering type-1 fuzzy relations in representing the *frequency of occurrence* and the *strength of confirmation*. Another proposal is the use of *bidirectional compound* binary fuzzy rules to represent medical knowledge without applying it to type-2 fuzzy set theoretical aspects and notations [27].

Regarding the use of conditional probabilities as multi-valued implications [28], the afore-mentioned study is similar to our model in terms of considering *conditional probabilities* as a type of inclusion relationship.

In the following, theoretical definitions for type-2 fuzzy set, type-2 fuzzy relation, and interval type-2 relations will be introduced on the basis of previous reports [16, 18, 27, 29–31], as preliminaries to the concept of interval type-2 fuzzy relation describing a fuzzy inclusion.

1.2 Preliminaries: Fuzzy Sets and Relations

Definition 1 (*Type-1 fuzzy set,* $\tilde{A}$)[4] A type-1 fuzzy set, denoted $\tilde{A}$ on the referential set $\mathcal{X} = \{x_1, x_2, ..., x_n\}$ is defined as a function $\mu_{\tilde{A}} : X \rightarrow [0, 1]$, i.e., as the set of pairs:

$$\tilde{A} = \{(x, \mu_{\tilde{A}}(x))|x \in \mathcal{X}\}. \tag{1}$$

This function is called a membership function. $\mu_{\tilde{A}}(x)$ is the degree of membership of the element $x \in \mathcal{X}$ in the set $\mu_{\tilde{A}}(x)$. Each membership degree $\mu_{\tilde{A}}(x)$ is *fully certain*, which means that in a type-1 fuzzy set, for each x value, there is no uncertainty associated with the primary membership value.

Definition 2 (*Type-2 fuzzy set,* $\tilde{\tilde{A}}$) Based on [29, 31], a type-2 fuzzy set denoted $\tilde{\tilde{A}}$, is defined as a function $\mu_{\tilde{\tilde{A}}} : X \times [0, 1] \rightarrow [0, 1]$, i.e., as the set of triples:

$$\tilde{\tilde{A}} = \{((x, u), \mu_{\tilde{\tilde{A}}}(x, u))|\forall x \in X, \forall u \in J_x \subseteq [0, 1]\}$$

here,

$$0 \leqslant \mu_{\tilde{\tilde{A}}}(x, u) \leqslant 1$$

For any given value of x, the $\mu_{\tilde{\tilde{A}}}(x, u), \forall u \in J_x$, is a type-1 membership function. J_x is used to reference the set of u values associated with each point in the X-axis.

---

[4] Based on [29, 31].

Based on Definition 2, a membership function of a type-2 fuzzy set can be represented by its graph in 2-D space[5]:

- The primary variable or the X-axis,
- the secondary variable or the Y-axis denoted by u, and
- the Z-axis, the membership function value (secondary grade); i.e. $\mu_{\tilde{\tilde{A}}}(x, u) \in [0, 1]$.

When uncertainties are removed, a type-2 membership function reduces to a type-1 membership function; i.e. the third dimension disappears.

To reference and describe the uncertainty in the primary memberships of a type-2 set, the concept of footprint of uncertainty or FOU is defined as:

Definition 3 (*Footprint of uncertainty, FOU*) Let $\tilde{\tilde{A}}$ be a type-2 fuzzy set, then:

$$FOU = \{\mu_{\tilde{\tilde{A}}}(x, u) | (x, u) \in X \times [0, 1]\}. \quad (2)$$

We can represent FOU as the union of all primary memberships:

$$FOU(\tilde{\tilde{A}}) = \bigcup_{x \in X} J_x$$

In Figs. 2 and 6, FOU is represented as shaded regions. Here, FOU is useful to access the minimal values and maximal values of uncertainties.

In our approach to simplify the complexity involved in a type-2 fuzzy set, the concepts of interval fuzzy set and interval type-2 fuzzy relation have been adopted to represent the uncertainty.

Definition 4 (*Interval type-2 fuzzy set*) A type-2 fuzzy set is an interval type-2 fuzzy set, if for every x there exists an interval $[\underline{u}, \overline{u}]$ such that $\mu(x, u) = 1$ for all u from this interval and $\mu(x, u) = 0$ for all other u.

Definition 5 (*Type-1 fuzzy relation,* $\tilde{R}$) Let $X = \{x_1, x_2, x_3, ..., x_n\}$ and $Y = \{y_1, y_2, y_3, ..., y_m\}$ be referential sets. A type-1 fuzzy relation, denoted $\tilde{R}$ on $X \times Y$, is defined as:

$$\tilde{R} : X \times Y \rightarrow [0, 1],$$

$$\tilde{R} = \{(x_i, y_j), \mu_{\tilde{R}}(x_i, y_j)\},$$

[5] For illustration see Fig. 1 in context of type-2 fuzzy relations.

with the membership function:

$$\mu_{\tilde{R}}(x_i, y_j) \in [0, 1].$$

For examples for type-1 fuzzy relations, see footnote 3.

Now we can introduce the concept of interval type-2 fuzzy relation. This concept is very important, as it will be employed in establishing uncertain and imprecise relationships between medical entities.

Definition 6 (*Interval type-2 fuzzy relation*) Let $X = \{x_1, x_2, x_3, ..., x_n\}$ and $Y = \{y_1, y_2, y_3, ..., y_m\}$ be referential sets. An interval type-2 fuzzy relation, denoted $\tilde{\tilde{R}}$ on $X \times Y$, is defined:

$$\tilde{\tilde{R}} : X \times Y \to \mathcal{F}([0, 1]),$$

where $\mathcal{F}([0, 1])$ represents the set of all subintervals of the interval [0,1]:

$$\mathcal{F}([0, 1]) = \{[x_L, x_U] : x_L,\ x_U \in [0, 1],\ x_L \leq x_U\},$$

$$\tilde{\tilde{R}} = \left\{(x_i, y_j), \left[\mu_{\tilde{R}}(x_i, y_j)_L, \mu_{\tilde{R}}(x_i, y_j)_U\right]\right\},$$

with the primary membership function:

$$\mu_{\tilde{R}}(x_i, y_j)_L, \mu_{\tilde{R}}(x_i, y_j)_U \in [0, 1]$$

and,

$$\forall (x_i, y_j), \mu_{\tilde{R}}(x_i, y_j)_L \leqslant \mu_{\tilde{R}}(x_i, y_j)_U$$

representing the lower and upper bound of $\tilde{\tilde{R}}$ elements respectively.

Based on Definitions 6 and 4, a type-2 fuzzy relation interval value is characterized by specific lower and upper boundaries instead of a fuzzy set, as is the case in type-2 fuzzy relations (Fig. 1). As all values of the secondary membership function equal 1, the uncertainty is represented by associated intervals.[6]

A type-2 fuzzy inclusion relation is characterized by a fuzzy set (Fig. 2).

---

[6] An example of an interval type-2 relation: s: "increased serum glucose" and d: "diabetes":
$$(s \xleftarrow[1]{} d, s \xrightarrow[[0.6,0.7]]{} d);$$
s always occurs with d but it only confirms d with certain possible values within [0.6,0.7].

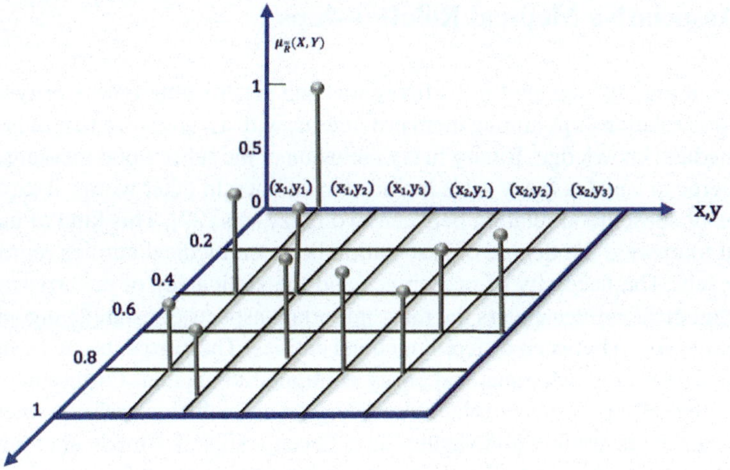

Fig. 1 Interval type-2 fuzzy relation (Definition 6)

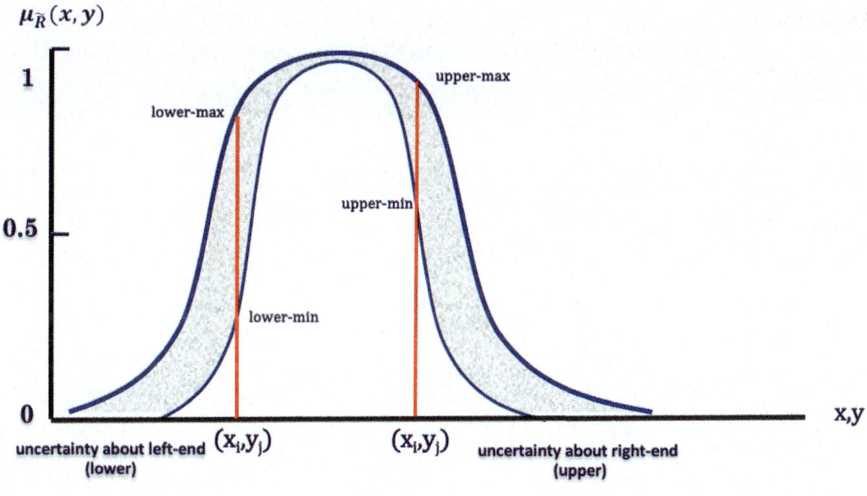

Fig. 2 Type-2 fuzzy relation. The footprint of uncertainty (FOU) is represented by the lower-min and upper-max of possible uncertain degrees of a type-2 fuzzy relation. Each end shows some uncertainties; we use lower-min and lower-max for the left-end, upper-min and upper-max for the right-end uncertainties

2 Associative Medical Relationships

Representing medical entities as fuzzy sets and establishing type-1 or type-2 fuzzy inclusion relationships among them provides us with a *framework* to represent uncertain medical knowledge. Binary fuzzy inclusion or the subsethood measure describes the degree to which a fuzzy set is included in another. In other words, it expresses the degree of subsethood relation between two fuzzy sets [27]. This kind of inclusion is useful to express the degree of association between medical entities represented as fuzzy sets. The necessity of occurrence and the sufficiency of occurrence between fuzzy medical entities covers the most important aspects of establishing an *associative relationship* between different medical entities. These aspects can be interpreted as the degree to which a medical entity is implied in another. Furthermore, considering interval type-2 fuzzy relations relying on an inclusion measure enables us to consider the uncertainty and vagueness between associative medical entities.

In the following we will present interval-valued fuzzy relations relying on the degree of subsethood to model the uncertainty and imprecision:

Definition 7 (*Type-1 fuzzy inclusion relation,* $\tilde{R}_I$) Let $\tilde{A}$ and $\tilde{B}$ be fuzzy subsets of $\mathcal{U} = \{x_1, x_2, ..., x_n\}$. A type-1 fuzzy inclusion relation, denoted $\tilde{R}_I$, is defined as

$$\tilde{R}_I : \mathcal{F}(\mathcal{U}) \times \mathcal{F}(\mathcal{U}) \to [0, 1],$$

where $\mathcal{F}(\mathcal{U})$ represents the set of all fuzzy sets in $\mathcal{U}$,

$$\tilde{R} = \left\{ \left((\tilde{A}, \tilde{B}), \mu_{\tilde{R}_I}(\tilde{A}, \tilde{B}) \right) \right\}$$

with

$$\mu_{\tilde{R}_I}\left(\tilde{A}, \tilde{B}\right) \triangleq \frac{\sum\limits_{x \in \mathcal{U}} \min\left(\mu_{\tilde{A}}(x), \mu_{\tilde{B}}(x)\right)}{\sum\limits_{x \in \mathcal{U}} \mu_{\tilde{A}}(x)} \in [0, 1] \; and \tag{3}$$

$$the\, scaler\, cardinality\, of\, A, |A| = \sum_{x \in \mathcal{U}} \mu_{\tilde{A}}(x) \neq 0.$$

Scalar inclusion measure expresses to which degree a fuzzy set is included in another one; i.e.

$$\mu_{\tilde{R}_I}(\tilde{A}, \tilde{B}) \triangleq degree(\tilde{A} \subseteq \tilde{B}) \in [0, 1] \tag{4}$$

and,

$$\mu_{\tilde{R}_I}(\tilde{B}, \tilde{A}) \triangleq degree(\tilde{B} \subseteq \tilde{A}) \in [0, 1] \tag{5}$$

These relations can be interpreted in terms of CADIAG-II as *strength of confirmation* (μ_c):

$$(s \underset{\mu_c}{\rightarrow} d),\ with\, \mu_c \in [0, 1]$$

and *frequency of occurrence* (μ_o):

$$(s \overset{\mu_o}{\leftarrow} d),\ with\, \mu_o \in [0, 1]$$

or *sufficiency* and *necessity* respectively.

Definition 8 (*Interval type-2 fuzzy inclusion relation,* $\tilde{\tilde{R}}_I$) The uncertainty of a type-2 fuzzy inclusion relation, denoted $\tilde{\tilde{R}}_I$, is associated with intervals given by type-1 fuzzy relation:

$$\tilde{\tilde{R}}_I : \mathcal{F}(\mathcal{U}) \times \mathcal{F}(\mathcal{U}) \rightarrow \mathcal{F}([0, 1]),$$

where $\mathcal{F}([0, 1])$ represents the set of all subintervals of the interval [0,1]:

$$\mathcal{F}([0, 1]) = \{[x_L, x_U] : x_L,\ x_U \in [0, 1],\ x_L \leq x_U\},$$

such that

$$\tilde{\tilde{R}}_I = \left\{\left((\tilde{A}, \tilde{B}), \left[\mu_{\tilde{\tilde{R}}_I}(\tilde{A}, \tilde{B})_L, \mu_{\tilde{\tilde{R}}_I}(\tilde{A}, \tilde{B})_U\right]\right)\right\},$$

$$\mu_{\tilde{\tilde{R}}_I}(\tilde{A}, \tilde{B})_L \leqslant \mu_{\tilde{\tilde{R}}_I}(\tilde{A}, \tilde{B})_U$$

The interval $[\mu_{\tilde{\tilde{R}}_I}(\tilde{A}, \tilde{B})_L, \mu_{\tilde{\tilde{R}}_I}(\tilde{A}, \tilde{B})_U]$ expresses the certain possible consistent degrees of the scalar inclusion relationship between $\tilde{A}$ and $\tilde{B}$, see Fig. 7.

Notably, a type-2 fuzzy inclusion relation interval value is characterized by specific lower and upper boundaries, while a *type-2 fuzzy inclusion relation* is characterized by a fuzzy set (Fig. 7 vs. Fig. 3).

Definition 9 (*Uncertain associative medical relationships*) Let $\mathcal{E} = \{e_1, e_2, e_3, ..., e_n\}$ be a set of medical entities represented as fuzzy sets. Uncertain associative relationship between medical entities can be interpreted as type-2 fuzzy inclusion relation.

The focus of this presentation will be on interval type-2 fuzzy relation, $\tilde{\tilde{R}}_I$:

$$\tilde{\tilde{R}}_I : \mathcal{F}(\mathcal{E}) \times \mathcal{F}(\mathcal{E}) \rightarrow \mathcal{F}([0, 1]).$$

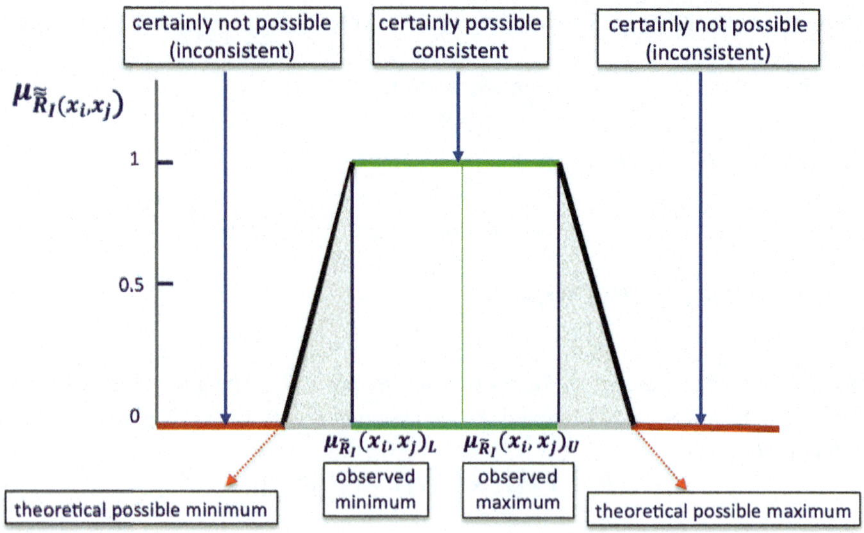

Fig. 3 General type-2 fuzzy inclusion relation. Values of type-1 fuzzy inclusion relation are characterized by type-2 inclusion relation. Values between observed minimum and maximum are certain

2.1 Acquiring Data for Associative Medical Relationships

Two major approaches may be used for dealing with medical knowledge acquisition, namely the classical knowledge-driven approach (symbolic representation in the context of linguistic uncertainty and imprecision), and the data-driven approach. The latter has been given greater importance in recent years, as we are approaching the era of big data and deep learning. Concrete data for instantiating associative medical relationships can be obtained from a variety of sources:

- Evaluating linguistic documentations by medical experts [13].
- Statistical analyses of medical patient databases [12].
- Data discovery in medical databases, i.e. utilizing data science methods on patient databases and documentations, such as predictive classification or descriptive methods (e.g., associative rule analysis), Fig. 4.

Domain experts cannot always deliver precise and consistent values for associative relationships without evaluating a *large quantity of medical data*. For example, a symptom that always occurs in a certain disease might *not be sufficient* to confirm the disease. One example of strong relationships would be *"highly increased amylase levels almost confirm acute pancreatitis"*. This type of associative relationship can be represented by considering a compound an interval type-2 fuzzy relation (see example in footnote 6).

The process of refinement of such intervals should be concluded by *checking* them for *consistency*. It should be noted that global consistency might refine the

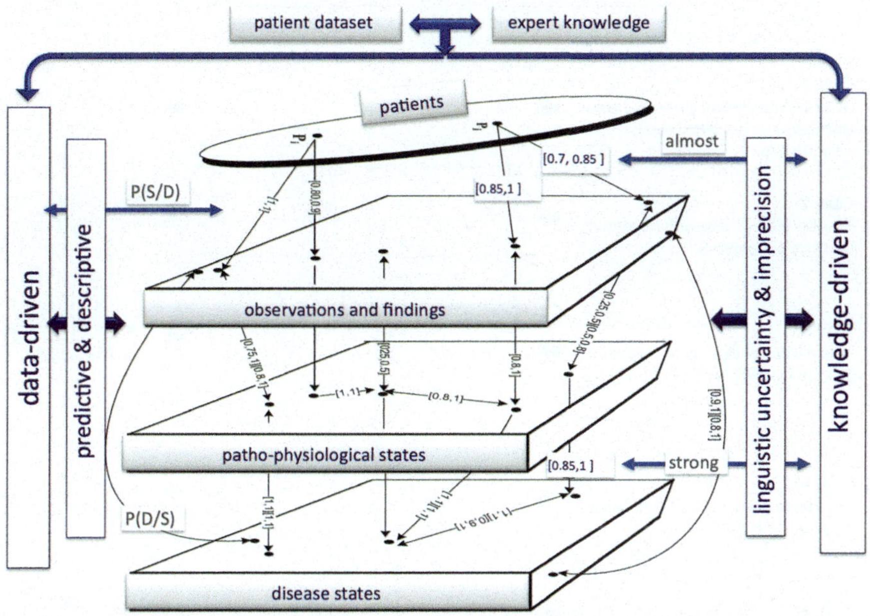

Fig. 4 Medical context of knowledge-driven and data-driven approaches for capturing uncertainty at the medical knowledge acquisition process. Interval-valued uncertainty is present at all levels of knowledge acquisition. In this Figure, the uncertain relationships such as (s $\xrightarrow[strong]{} d$) can be expressed in terms of interval type-2 fuzzy relation, formalized as (s $\xrightarrow[[0.85,1]]{} d$)

upper and *lower* values, so that useful global *minima* might be acquired. Establishing such uncertainties requires a *stepwise knowledge acquisition* process and refinement; some cases are provided in Fig. 5. In such processes, an associative relationship might start with no prior knowledge or may be a simple association, such as a *positive or negative correlation*, and might end with a type-1 fuzzy relation, (Definition 1, $\tilde{R}_I$) or a *consistent interval* (Definition 8, $\tilde{\tilde{R}}_I$). In each step or phase, the expert may add knowledge that would refine the degree of imprecision and uncertainty. However, as associative relationships in all phases might be affected by some degree of uncertainty, useful inferential knowledge can be successively added to the acquisition process.

Furthermore, *initial values* can be estimated *statistically* by analyzing a medical database. This approach has been successfully employed as semi-automatic knowledge acquisition within the knowledge-based system CADIAG-II/Rheuma [3, 12]. In this context, necessity (*frequency of occurrence*) may be interpreted as P(S/D), and sufficiency (*strength of confirmation*) as P(D/S), which might be estimated via Bayes' theorem and refined or transformed to fuzzy values.

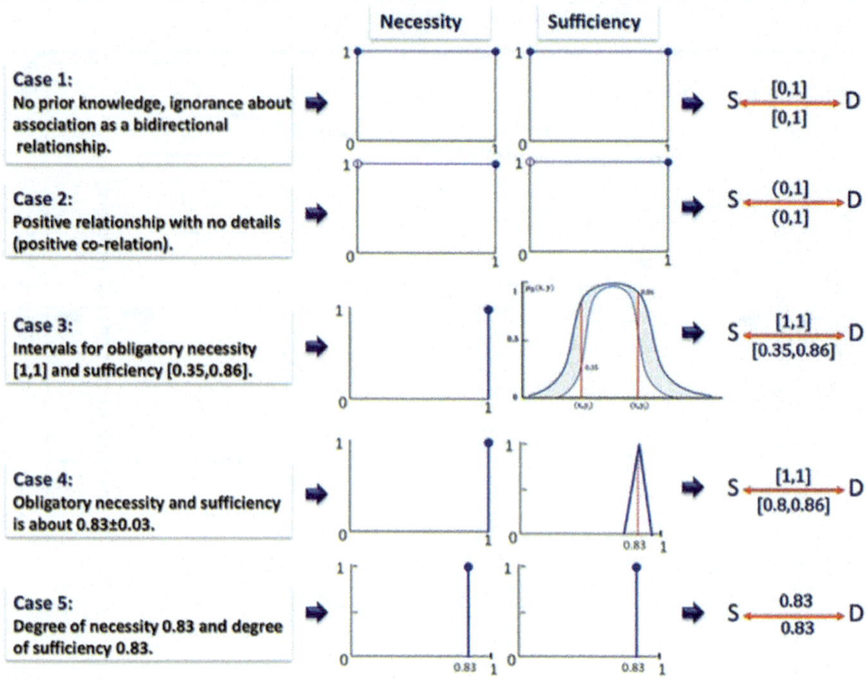

Fig. 5 Example of representing medical uncertainty in the context of necessity and sufficiency of occurrence between medical entities based on the concept of interval type-2 fuzzy relation. The boundaries of FOU, footprint of uncertainty, can be reduced by a refinement process checking the boundaries for local and global consistency

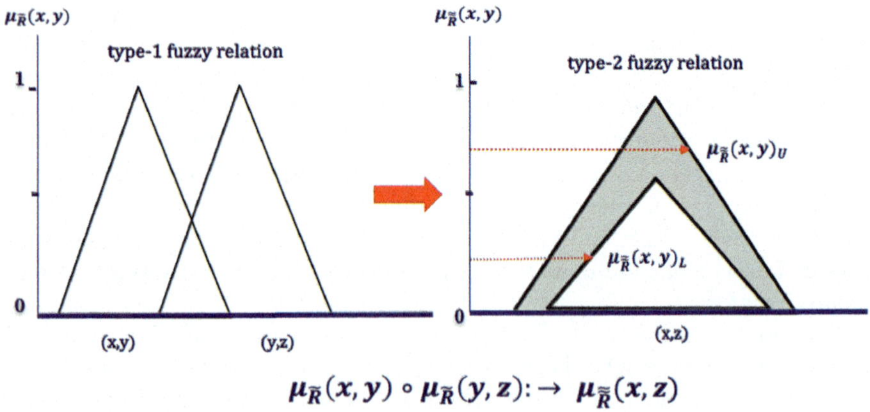

Fig. 6 The composition of type-1 binary fuzzy relations results in a type-2 relation when enhancing it with uncertainty by a local triangular dataset representation

3 Inferencing Type-2 Fuzzy Relation

As mentioned earlier, a variety of sources may be used to acquire concrete data for instantiating associative medical relationships, such as evaluating linguistic documentations, statistical analysis of medical patient databases, or data-driven tasks. However, the creation of knowledge bases with a large number of relationships between medical entities might result in inconsistencies and incompleteness. Furthermore, in many cases, decision-making under imprecision and uncertainty is required.

Several human domain experts might suggest inconsistent estimations of associative relationships in the context of *relevance estimation and assessment*. In some cases, an *agreement* or *disagreement analysis* of the involved experts should be considered. The grade of agreement or disagreement or bias might be used as a reference for considering the degree of uncertainty [14].

To access this important aspect, we need an inferential model that is capable of computing all possible consistent values for a type-1 and even type-2 fuzzy relation. Values lying outside these intervals should be considered as inconsistent values (Fig. 7). Systems affected by inconsistency might reduce the performance of a knowledge-based system.

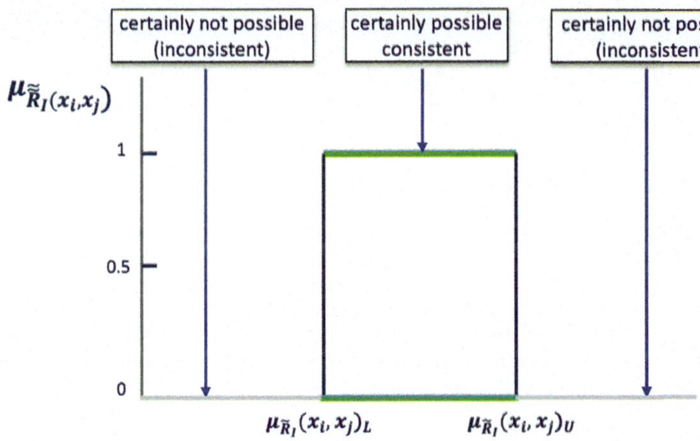

Fig. 7 Interval type-2 fuzzy relation representing an interval-valued binary fuzzy relationship. The certain possible values with $\mu_{\tilde{\tilde{R}}_I}(x_i, y_j) = 1$ are consistent. The final goal is to compute the interval of such type-2 inclusion fuzzy relations

3.1 Inferential Uncertain Relations

This illustration (Fig. 7) extends from the inconsistent interval of uncertainty to certainly not possible values for uncertainties. The present paper will be limited to introducing the *basic concept* of the *inferential model* in the case of type-1 fuzzy relation.[7] This type has been widely used in the different implementations of CADIAG-II-based knowledge-based systems [5–11].

Notably, the computed consistent intervals might be interpreted as boundaries for possible uncertainties arising from assuming the certainty of precise values such as $\mu_{\tilde{A}}(x)$. Figure 6 shows that the composition of two type-1 fuzzy relations, i.e. certain fuzzy relations, would propagate uncertainty in form of consistent intervals.[8]

In the following, we will focus on the basic case of inferring consistent intervals within locally investigated triangular datasets to infer consistent intervals and their minima for the *upper* and *lower* boundaries; i.e. FOU. In this model, we differentiate between local and global uncertainty.

Let $\mathcal{M}$ be a triangular dataset of medical entities consisting of point-valued type-1 relations (Definitions 5, 7), $\tilde{R}_{\mathcal{I}}$:

$$\begin{aligned}\mathcal{M} = \{e_1 \xrightarrow[a_2]{} e_2, e_1 \xleftarrow{a_1} e_2,\\ e_2 \xrightarrow[b_2]{} e_3,\\ e_2 \xleftarrow{b_1} e_3\}\end{aligned}$$

The possible consistent type-1 fuzzy relationships between e_1 and e_3 can be computed as *interval type-2 relation*, i.e.:

$$\mathcal{M} \Vdash \{e_1 \xrightarrow[[\underline{x}_1, \bar{x}_1]]{} e_3, e_3 \xleftarrow[[\underline{x}_2, \bar{x}_2]]{} e_1\}$$

The interval $[\underline{x}_i, \bar{x}_i]$ represents the lower and upper boundaries for uncertainty; i.e.:

$$\begin{aligned}\underline{x}_1 &= \mu_{\tilde{R}_I}(e_1, e_3)_L\\ \bar{x}_1 &= \mu_{\tilde{R}_I}(e_1, e_3)_U\\ \underline{x}_2 &= \mu_{\tilde{R}_I}(e_3, e_1)_L\\ \bar{x}_2 &= \mu_{\tilde{R}_I}(e_3, e_1)_U,\end{aligned}$$

[7] Considering all other aspects, such type-2 inference exceeds the scope of the current presentation.

[8] Obligatory sufficiency and necessity of yield certainty, refer to Fig. 6, i.e.:

$$\mu_{\tilde{R}_{\mathcal{I}}}(x, y) = 1, \mu_{\tilde{R}_{\mathcal{I}}}(y, x) = 1, \mu_{\tilde{R}_{\mathcal{I}}}(y, z) = 1, \text{ and } \mu_{\tilde{R}_{\mathcal{I}}}(z, y) = 1$$

and they are computed as follows:

$$\underline{x}_1 = \begin{cases} a_2 - \dfrac{a_2}{a_1} \cdot (1 - b_2) & \text{if } b_2 > 1 - a_1, a_1 \neq 0 \\ 0 & otherwise \end{cases} \tag{6}$$

$$\bar{x}_1 = \begin{cases} \min\left(a_2, \left(\dfrac{b_2 \cdot a_2}{a_1}\right)\right) + \min\left(1 - a_2, \left(\left(\dfrac{b_2 \cdot a_2}{b_1 \cdot a_1}\right) \cdot (1 - b_1)\right)\right) & \\ \quad \text{if } b_1 \neq 0, a_1 \neq 0 & \\ 0 & \text{if } b_1 = 1, a_1 = a_2 = 0 \\ 1 - a_2 & otherwise \end{cases} \tag{7}$$

$$\underline{x}_2 = \begin{cases} \underline{x}_1 \cdot \left(\dfrac{b_1 \cdot a_1}{b_2 \cdot a_2}\right) & \text{if } b_2 \neq 0, a_2 \neq 0 \\ 0 & otherwise \end{cases} \tag{8}$$

$$\bar{x}_2 = \begin{cases} \bar{x}_1 \cdot \left(\dfrac{b_1 \cdot a_1}{b_2 \cdot a_2}\right) & \text{if } b_2 \neq 0, a_2 \neq 0 \\ 0 & \text{if } b_1 = 0, b_2 = 0, a_1 = 1 \\ 1 - b_1 & otherwise \end{cases} \tag{9}$$

The derivation of these formulae can be achieved by considering all possible inclusion degrees within the triangular dataset $\mathcal{M}$; such as $degree(e_1 \subset e_2)$, $degree(e_2 \subset e_1)$, $degree(e_2 \subset e_3)$, and $degree(e_3 \subset e_2)$ in context of computing the minimal and maximal $degree(e_1 \subset e_3)$, and $degree(e_3 \subset e_1)$. All these relationships can be expressed in terms of constraints represented as computable linear equalities and/or inequalities. Solving all these constraints in reference to $|e_1| = \sum_{x \in \mathcal{U}} \mu_{e_1}(x)$ yields an interval of possible degrees for $(e_1 \supset e_3)$ and $(e_3 \supset e_1)$. The basic idea of the derivation can also be found in [27].

Example 1 Let $\mathcal{M}$ be a triangular set of relations of a type-1 fuzzy relation; Definition 5:

$$\mathcal{M} = \{e_1 \xrightarrow[0.5]{} e_2, e_1 \xleftarrow{0.75} e_2,$$
$$e_2 \xrightarrow[1]{} e_3,$$
$$e_2 \xleftarrow{0.25} e_3\}$$

Based on the Eqs. 6 and 7, all instances of possible relationships are:

$$(e_1 \xrightarrow[{[\underline{x}_1, \bar{x}_1]}]{} e_3) \in \mathcal{S},$$

where $\mathcal{S}$:

$$\mathcal{S} = \{(e_1 \xrightarrow[{[\underline{x}_1, \bar{x}_1]}]{} e_3) | [\underline{x}, \bar{x}] \subseteq [0.5, 1]\} \tag{10}$$

are consistent instances with $\mathcal{M}$ in terms of uncertainty.

The computation of globally consistent intervals requires inference of minimal intervals globally, which can be implemented *incrementally* and *recursively*.

3.2 Application Potential

As mentioned earlier, this approach is founded on the following aspects:

- Employing interval type-1 and type-2 fuzzy relations expressing the necessity and sufficiency of occurrence in the context of establishing associative relationships between medical entities; e.g., point-valued, linguistic and interval-valued:

$$(s \xrightarrow[\mu]{} d, s \xrightarrow[strong]{} d, s \xrightarrow[{[a,b]}]{} d)$$

  The basic concepts were *partly* employed in designing a CADIAG-II-based system such as MedFrame/CADIAG-IV, and implemented as a stepwise incremental refinement acquisition system [13].

- Inferencing useful consistent intervals to refine and possibly derive new associative relationships, and to check the knowledge base for logical inconsistencies.[9]
- Finally, integrating the compositional rule of inference within this model in connection with an inference engine within a decision support system.

The following example illustrates the application potential of integrating the inference model into refinement and data quality assurance:

Example 2 Let $\mathcal{M}$ be a triangular dataset of compound relationships as defined in Example 1. Based on 3.1, the inferred relationships:

$$(e_1 \xrightarrow[{[0.5,1]}]{} e_3)$$

and

[9] The authors are working on integrating the introduced inference model within the stepwise refinement process. However, further research will be needed to consider complex relationships expressing logical combinations of medical entities on the left side of a rule in the context of the global and local consistency.

$$(e_1 \xleftarrow[{[0.187,0.375]}]{} e_3)$$

can be represented as a consistent bi-directional (compound) interval-valued type-2 relationship, (Definition 8)

$$e_1 \overset{[0.187,0.375]}{\underset{[0.50,1]}{\longleftrightarrow}} e_3. \quad (11)$$

Such relationships are very useful for the following tasks:

- Checking an relationship for logical consistency (10). For instance, the rule in (12) describes a possible relationship expressing the degree of sufficiency and it is consistent with $\mathcal{M}$ with some certainly possible values:

$$e_1 \xrightarrow[{[0.75,0.95]}]{} e_3, \quad (12)$$

  while the relationship in (13):

$$e_1 \xrightarrow[{[0.25,0.345]}]{} e_3 \quad (13)$$

  represents an inconsistent relationship, as the values exceed the scope of the certainly possible values, see Fig. 7.
- The relationship between e_1 and e_3 in (11); e.g., $e_1 \xrightarrow[{[0.50,1]}]{} e_3$, and $(e_1 \xleftarrow[{[0.187,0.375]}]{} e_3)$ can be added to the knowledge base to increase some issues related to performance and completeness.
- Under the assumption that previous knowledge has already been validated on consistency, this approach relies on consistent interval propagation. In case an expert would propose new values for $(e_i \rightarrow e_j)$, the new values are expected to lie within the certainly possible values of the computed formula. However, further interval refinement is possible by considering new knowledge. The refinement process (i.e. narrowing the fingerprint of certainty) can be achieved by computing the global consistency of the model under the newly added values.
- Finally, integrating the compositional rule of inference within this inference, we can follow an inference engine within a decision support system.

4 Conclusion and Future Perspectives

This paper describes the handling of some crucial aspects of knowledge representation, relying on the acquisition of consistent inferential associative relationships. The adopted approach emphasizes the importance of considering fuzzy sets and type-2 relations with a view to the establishment of associative medical relationships within an inferential model, considering uncertainty, and checking for logical consistency. Many aspects of this model have been successfully employed by different implementations of CADIAG-II-based systems. For future work, the integration of this approach within a stepwise refinement of the knowledge acquisition process will be significant for ensuring data quality and enhancing performance. Furthermore, an interval-based compositional rule of inference might lead to a form of reasoning that relies on inferencing consistent uncertain intervals. Finally, it would be desirable to fine-tune the established intervals by integrating data-driven approaches. Clustering, relationships, and interval-valued-based deep learning might be useful approaches.

Acknowledgements We are indebted to Andrea Rappelsberger for her extended assistance in formatting and finalizing this paper.

References

1. K.-P. Adlassnig, G. Kolarz, W. Scheithauer, H. Effenberger, G. Grabner, CADIAG: approaches to computer-assisted medical diagnosis. Comput. Biol. Med. **15**(5), 315–335 (1985)
2. K.-P. Adlassnig, G. Kolarz, Representation and semiautomatic acquisition of medical knowledge in CADIAG-1 and CADIAG-2. Comput. Biomed. Res. **19**(1), 63–79 (1986)
3. G. Kolarz, K.-P. Adlassnig, Problems in establishing the medical expert systems CADIAG-1 and CADIAG-2 in rheumatology. J. Med. Syst. **10**(4), 395–405 (1986)
4. W. Moser, K.-P. Adlassnig, Consistency checking of binary categorical relationships in a medical knowledge base. Artif. Intell. Med. **4**(5), 389–407 (1992)
5. K.-P. Adlassnig, A fuzzy logical model of computer-assisted medical diagnosis. Methods Inf. Med. **19**(3), 141–148 (1980)
6. K.-P. Adlassnig, Fuzzy set theory in medical diagnosis. IEEE Trans. Syst. Man Cybern. **16**(3), 260–265 (1986)
7. K.-P. Adlassnig, Uniform representation of vagueness and imprecision in Patient's medical findings using fuzzy sets, in *Cybernetics and Systems '88*, ed. by R. Trappl (Kluwer Academic Publishers, Dordrecht, 1988), pp. 685–692
8. K.-P. Adlassnig, M. Akhavan-Heidari, CADIAG-2/GALL: an experimental expert system for the diagnosis of gallbladder and biliary tract diseases. Artif. Intell. Med. **1**(2), 71–77 (1989)
9. K.-P. Adlassnig, H. Leitich, G. Kolarz, On the applicability of diagnostic criteria for the diagnosis of rheumatoid arthritis in an expert system. Expert Syst. Appl. **6**(4), 441–448 (1993)
10. H. Leitich, K.-P. Adlassnig, G. Kolarz, Development and evaluation of fuzzy criteria for the diagnosis of rheumatoid arthritis. Methods Inf. Med. **35**(4–5), 334–342 (1996)
11. H. Leitich, P. Kiener, G. Kolarz, C. Schuh, W. Graninger, K.-P. Adlassnig, Prospective evaluation of the medical consultation system CADIAG-II/RHEUMA in a rheumatological outpatient clinic. Methods Inf. Med. **40**(3), 213–220 (2001)
12. H. Leitich, K.-P. Adlassnig, G. Kolarz, Evaluation of two different models of semi-automatic knowledge acquisition for the medical consultant system CADIAG-II/RHEUMA. Artif. Intell. Med. **25**(3), 215–225 (2002)
13. K. Bögl, K.-P. Adlassnig, Y. Hayashi, T.E. Rothenfluh, H. Leitich, Knowledge acquisition in fuzzy knowledge representation framework of a medical consultation system. Artif. Intell. Med. **30**(1), 1–26 (2004)
14. B. Haddad, Relevance & assessment: cognitively motivated approach toward assessor-centric query-topic relevance model. Acta Polytechnica Hungarica **15**(5), 129–148 (2018)

15. L.A. Zadeh, Outline of a new approach to the analysis of complex systems and decision process. IEEE Trans. Syst. Man Cybern. **3**(1), 28–44 (1973)
16. N.N. Karnik, J.M. Mendel, Q. Liang, Type-2 fuzzy logic systems. IEEE Trans. Fuzzy Syst. **7**(6), 255–269 (1999)
17. J.M. Mendel, *Uncertain Rule-Based Fuzzy Logic Systems: Introduction and New Directions* (Prentice-Hall Inc, Upper Saddle River, 2001)
18. R. John, J. Mendel, J. Carter, The extended sup-star composition for type-2 fuzzy sets made simple, in: *2006 IEEE International Conference on Fuzzy Systems, Sheraton Vancouver Wall Centre Hotel, Vancouver, BC, Canada, July 16–21, 2006* (2006), pp. 1441–1445
19. S. Liu, R. Cai, Uncertainty of interval type-2 fuzzy sets based on fuzzy belief entropy. Entropy **23**, 1265 (2021)
20. J.F. Baldwin, Support logic programming. Int. J. Intell. Syst. **1**(2), 73–104 (1986)
21. I.B. Turksen, Interval-valued fuzzy sets and compensatory 'AND'. Fuzzy Sets Syst. **51**(3), 295–307 (1992)
22. J. Fan, W. Xie, J. Pei, Subsethood measures: news. Fuzzy Sets Syst. **106**(2), 201–209 (1999)
23. B. Kosko, *Neural Networks and Fuzzy Systems: A Dynamical Systems Approach to Machine Intelligence* (Prentice-Hall Inc, Upper Saddle River, 1991)
24. V.R. Young, Fuzzy subsethood. Fuzzy Sets Syst. **77**, 371–384 (1996)
25. B. Haddad, Cognitively motivated query abstraction model based on root-pattern associative networks. J. Intell. Syst. **29**(1), 910–923 (2020)
26. C.M. Helgason, T.H. Jobe, Perception-based reasoning and fuzzy cardinality provide direct measures of causality sensitive to initial conditions in the individual patient. Int J Comput Cognit **1**(2), 79–104 (2003)
27. B. Haddad, A. Awwad, Representing uncertainty in medical knowledge: an interval based approach for binary fuzzy relations. Int Arab J Inf Technol **7**(1), 63–69 (2010)
28. J. Heinsohn, B. Owsniki-Klewe, Probabilistic inheritance and reasoning in hybrid knowledge representation systems, in W. Hoeppner (Hrsg.), *Künstliche Intelligenz, Proceedings GWAI-88, 12. Jahrestagung, Eringerfeld, 19.–23. September 1988, Informatik-Fachberichte 181* (Springer, Berlin, 1988), pp. 51–60
29. J.M. Mendel, Type-2 fuzzy sets: some questions and answers, in *IEEE Neural Networks Society Newsletter August 2003*, vol. 1 (2003), pp. 10–13
30. Q. Liang, J.M. Mendel, Interval type-2 fuzzy logic systems: theory and design. IEEE Trans. Fuzzy Syst. **8**(5), 535–550 (2000)
31. J.M. Mendel, R.I.B. John, Type-2 fuzzy sets made simple. IEEE Trans. Fuzzy Syst. **10**(2), 117–127 (2002)

Mental States Detection by Extreme Gradient Boosting and k-Means

Nam Anh Dao and Quynh Anh Nguyen

Abstract We present a categorical Electroencephalogram (EEG) detection model to study stress and predict the cognitive states among crew members. A two-stage, learning is addressed: a unsupervised clustering is applied firstly for training dataset and the clusters' centers are used for clustering test dataset; the second stage implements gradient-boosted decision tree for classifying mental states. The mode requires data analysis for representing the EEG signals in the best possible minimal storage space and freeing from noise so the data can be clear and well shaped. While the PCA interpreted the EEG features by principal components and reduces the dimension of the data, the autoencoder yielded significant representation for the data ignoring unimportant data. We show competitive results and discuss efficiency of combining supervised and unsupervised learning for crew members' EEG classification.

Keywords Electroencephalogram · k-means · PCA · Autoencoder · XGBoost · Fly safety

1 Introduction

Several recent works on the EEG have explored the aspects of both clinical diagnosis and scientific research. Actually, a particular EEG signals [1] are the electrical currents that are produced by cortical neurons and measured at scalp. As a specific type of bio-signals with high time-resolution and the adequate accuracy, EEG can be explored by statistical analysis and machine learning. In this way, metal status like emotion, stress and mental workload, can be automatically detected from learning models. We face the question of analysis of Electroencephalogram (EEG) for checking mental states of crew members. The target is to capture specific patterns from EEG signals for automatic identification of mental states in new signals, where detecting mental states of crew members can support controlling aviator safety [2]. The reason

N. A. Dao (✉) · Q. A. Nguyen
Electric Power University, Hanoi, Vietnam
e-mail: anhdn@epu.edu.vn

N. H. Phuong and V. Kreinovich (eds.), *Deep Learning and Other Soft Computing Techniques*, Studies in Computational Intelligence 1097,
https://doi.org/10.1007/978-3-031-29447-1_2

of learning mental states is that personal physical and emotional state, ability and motivation are related to individual task performance [3]. More specifically, alerting workload could minimize the risk of unsafe driving [4].

In this article, we specify a method that can assist EEG analysis with classification regarding the clustering of the samples. The unsupervised learning task is at the heart of the mental state classification method for crew members that we demonstrate in the third section. The method have essential functions: (a) careful data analysis is supported by signals representation in restraint memory and noise removal; (b) constructive features are extracted by principal component analysis (PCA) [5] and autoencoder [6] reducing the dimension of data; (c) clusters and centroids are determined by clustering training data, allowing test data clustered by the centroids; (d) supervised classification is performed for each cluster by implementing the extreme gradient boosting. As the results, mental states are predicted with improved performance.

The article is organized as follows: in Sect. 2 we review related papers and show the achievements in the EEG domain. We describe our method and the ideas underlying the our use of mixing learning techniques in the Sect. 3. Demonstration of the method is shown in the Sect. 4 by experiments performed with support of a benchmark database.

2 Related Works

Let us review important published applications and solutions concerning EEG signals and their relation with the pilots' workloads.

A thesis by Crijnen [7] provided analysis of frequency domain and by sliding window on the time axis. Actually, the work performed cognitive state classification and detection of the change of cognitive states by using time series techniques. In relation to reducing commercial aircraft fatalities, Mishra et al. [8] implemented a machine learning model based on SVM as a classifier with a real physiological data. The database is published by the Kaggle [9] covering EEG data and cognitive states of 400 pilots for predicting mental states of pilots during flight. Also, a classification model with the same dataset from the Kaggle [9] is proposed by Lin [10] using the some methods to detect pilots' mental states as either "safe" or "dangerous". The model interpreted the time series data using several models including Logistic Regression, Support Vector Machine (SVM), k-Nearest Neighbors(KNN), Recursive Partitioning and Regression Trees, Random Forests. This method achieved an accuracy of approximately 90% reliability to distinguish pilots' cognitive states.

This is evidence, a pattern classification method using extreme gradient boosting, random forest and support vector machine classifiers was introduced by Harrivel et al. [11] for performing multimodal classification. This model gets pre-processed electroencephalography, galvanic skin response, electrocardiogram, and respiration signals as input features for predicting cognitive states during flight simulation.

Hernández-Sabaté et al. [12] considered the convolutional neural network as an effective classifier to link EEG features with divergent mental workloads. The work performed experiment that partially quantifies memory capacity and working memory. Thus, Binias et al. [13] organized an experiment that simulated aircraft flights with 19 participants in a 2-h long session. EEG data of the pilots was used for prediction algorithms that include Kernel Ridge and Radial Basis Support Vector Machine regression as well as the Least Absolute Shrinkage Operator and its Least Angle Regression modification.

It is possible to construct systems for predicting metal states of pilots having EEG signals by mentioned above learning models like SVM, Naïve Bayes and convolutional neural network. However, this classification is challenged because of extremely large and complex data as the EEG data is time series signals from multiscalp sites for each person. For these reasons we propose a model combining different learning models for achieving efficient performance in both accuracy and computing time.

3 The Method

In what follows, we describe a learning model for studying stress predicting cognitive states in the group of crew members. The goal of registration of EEG signals extracted from crew members to their mental states is to find a transformation which maps any EEG signal x in X to its corresponding mental state y in Y. As will be seen, the learning model performs three stages: data analysis, feature engineering as well as learning.

In the first stage of data analysis, the EEG data is studied to perform data type conversion, so that the best possible minimal storage space of the data is achieved. This task allows reducing data storage and improving computing speed (see arrow 2 in Fig. 1). In practice, however, noise is discovered in the EEG signals, and so, in this stage, the noise is removed to get clear signals (arrow 3 in Fig. 1).

In the following stage we first select the principal component analysis (PCA) features according to how well they interpret the EEG signals, and to reduce the dimension of the data (arrow 5 in Fig. 1). By applying autoencoder for EEG data, a significant representation for the data is generated and dimensionality is also reduced to ignore unimportant data (arrow 6 in Fig. 1). In fact, the PCA and autoencoder permit to extract features from the EEG data for further learning task.

During the first step in the learning stage, both training and test data set are defined by splitting the EEG data with described above features (see arrow 9 in Fig. 2). The resulting training data set, which can have any number of labeled mental states, is then clustered into segments using k-means, then centroids of the segments are calculated and saved (arrow 11 in Fig. 2). The splitting task is processed with multi folds for cross validation.

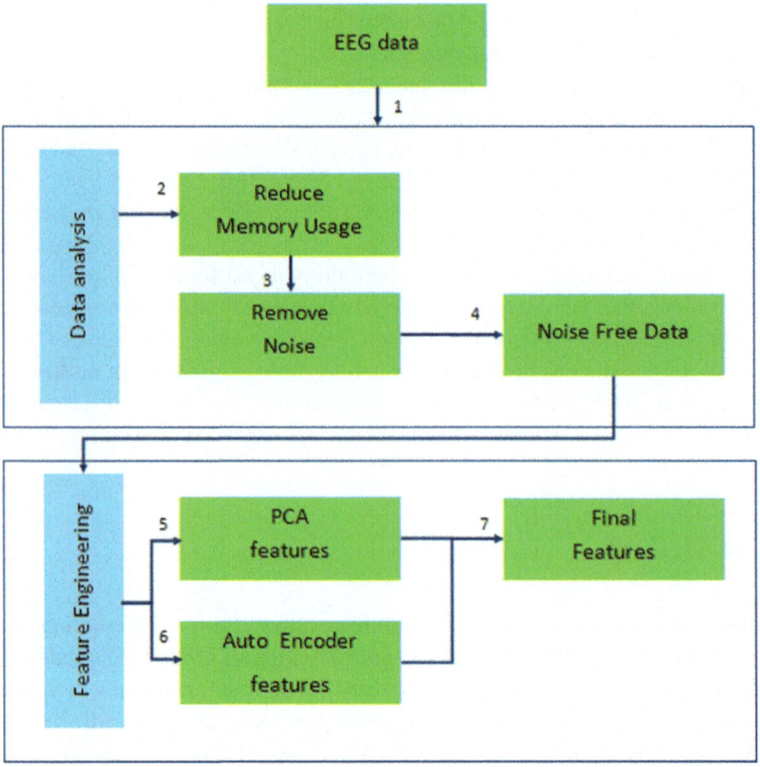

Fig. 1 Data analysis and feature engineering

To create segments for the test dataset we used the centroids and distance function for estimating distance from a test sample to each centroid (arrow 15, 16 in Fig. 2). With Extreme Gradient Boosting (XGBoost) [14], which is a scalable, distributed gradient-boosted decision tree, the training dataset of predefined above segment is used for training the learning model (arrow 13, 14 in Fig. 2). Hence, followed test set of the same segment with the trained model can be classified into metal states (arrow 17, 18 in Fig.2).

The truth is, of course, that cross validation by performing several folders of each test case are important to validate proposed model. Thus, performance metrics can be measured in each split and collected for getting final averaged values. Our metrics which are suitable for our article covers accuracy, precision and f1-score. Clearly, the formulas or the metrics are as follows where n_{TP}, n_{TN}, n_{FP} and n_{FN} are the number of true positive samples, true negative samples, false positive samples and false negative samples, respectively:

$$accuracy = \frac{n_{TP} + n_{TN}}{n_{TP} + n_{FP} + n_{TP} + n_{FN}}, \quad (1)$$

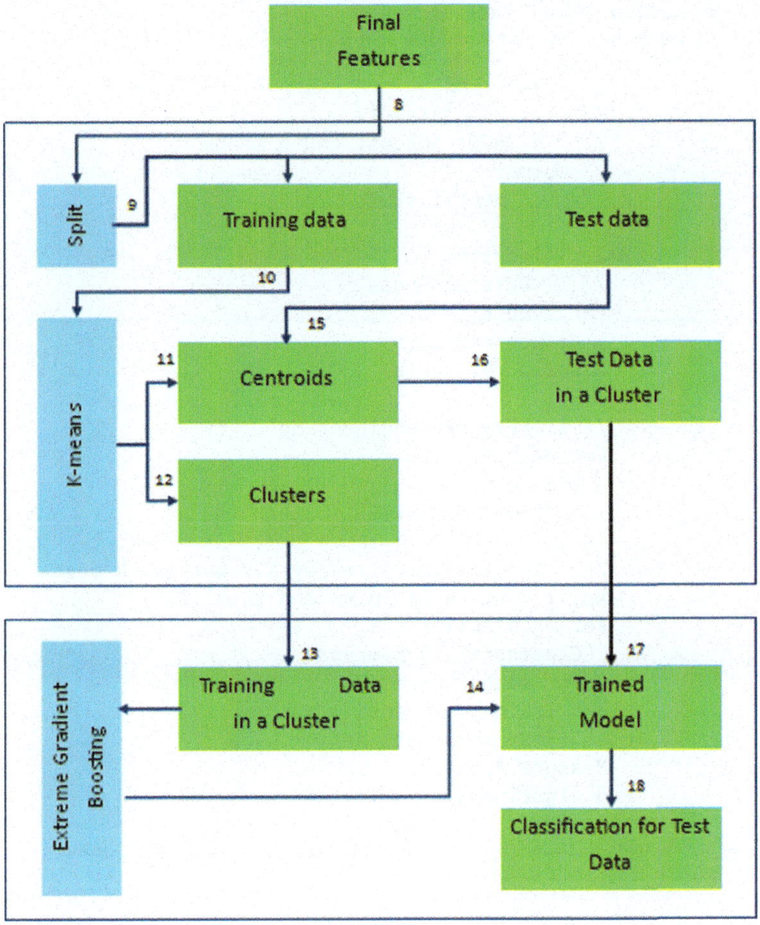

Fig. 2 Learning process

$$precision = \frac{n_{TP}}{n_{TP} + n_{FP}}, \tag{2}$$

$$f1\_score = \frac{2 \times recall * precision}{recall + precision}. \tag{3}$$

4 Experiments

Given this picture of described learning model for classifying EEG signals to mental states, where is data and what is actual performance of tests. In a project like this, we will describe in details in four subsections: data analysis, feature engineering, learning procedure and performance report (see Fig. 3).

Algorithm. Mental States Detection by Extreme Gradient Boosting and k-means

```
    // Input:
    //   e: EEG dataset
    //   METHOD: XGBoost and k-means
    // Output:
    //   s: mental state

Start Algorithm
    // I. Data analysis
    For each e:
        Reduce memory(e)
        Noise removal(e)

    // II. Feature engineering
    For each e:
        PCA (e)
        Auto-encoder (e)

    // III. Classification by XGBoost and k-means
    Split data to training and test sets
    For each split:
        Clustering by k-means
        For each clusrer:
            Train by XGBoost
            Test with test set
    Measure Accuracy
End Algorithm
```

Fig. 3 Algorithm mental states detection by extreme gradient boosting and k-means

A. Data analysis. The story begins when the Kaggle organized a prediction competition which is called Reducing Commercial Aviation Fatalities [9]. There was primarily a challenge for detecting troubling events from aircrew's EEG dataset. We use this dataset for our experiments as it provided real physiological data from eighteen pilots who were subjected to various distracting events. The dataset is collected in a non-flight environment and outside of a flight simulator.

Description of the data is shown in Table 1. You will see that, the physiological data includes EEG, ECG, Respiration and Galvanic skin response signals. The dataset contains 28 features where the crew feature is an unique id for a pair of pilots and there are total 9 crews.

It is widely accepted that pilots are evaluated with three statuses: the Chanenezd attention (CA) which is the state in which the pilot is always focused on a specific task without regard to other missions; the Diverted attention (DA) that is a pilot's state of being diverted from one mission to another when there is action or thought involved; and the Startle/Surprise (SS), that is appeased by having the subjects like watching movie clips with jump scares.

Table 1 Data structure in experiments

Column	Description
Crew	ID for a pair of pilots
Experiment	CA—channelized attention; DA—Diverted attention; SS—startle/Surprise
Subject Amount	18
Seat	0 (left); 1(right)
EEG	eeg_fp1, eeg_f7, eeg_f8, eeg_t4, eeg_t6, eeg_t5, eeg_t3, eeg_fp2, eeg_o1, eeg_p3, eeg_pz, eeg_f3, eeg_fz, eeg_f4, eeg_c4, eeg_p4, eeg_poz, eeg_c3, eeg_cz, eeg_o2
ECG	3-point Electrocardiogram signal
GRS	Galvanic Skin Response
R	Measure of the rise and fall of the chest
Event	Baseline, CA, DA, SS

Next, the seat feature is the pilot in the left or right seat. Here, electroencephalogram (EEG) signals have 20 features. There are some other features: ECG—Electrocardiogram signal feature; R—Respiration feature which is a measure of the rise and fall of the chest; GRS—Galvanic Skin Response that is a measure of electrodermal activity. Event feature is the state of the pilot at the given time. It is one of four categorical values: A for baseline, B for SS, C for CA and D for DA.

B. Feature Engineering. We will explore three steps of the pre-processing. The first step is for reducing memory and noise removal. The second step is exploring train data and the last step is for feature extraction. We modified the data type to reduce memory usage from 1039.79 MB to 241.38 MB achieving 76% then removed the noisy data from specific two columns: ECG and R. Considering EEG features with 20 columns, we reduced them to 2 components by the principal component analysis (PCA).

Once noise free EEG features has been created, it is the time for learning a representation (encoding) for the data by auto-encoder, which performs training the network to ignore insignificant data. So that dimension is then reduced. When the training is completed, the decoder provides hidden features which can be used for further learning. Note, as an indispensable task, the data is standardized by transforming all features which are in different scale and types into a consistent format that adheres to the standards.

C. Learning Process. For the purposes of cross validation, our data is split into train and test set in 5 folds. Here, for each fold, we conducted clustering by k-means for training data resulting clusters and centroids. In particular, the centroids help to assign appropriate clusters for test data. In this study, several number of clusters was applied for the clustering, including 3,4,5,6,7,8 for each fold.

We will now look in turn at each of the cluster where training and test sets are ready. In each fold, the learning model of XGBoot is implemented for predicting pilot's mental state which belong to one of four classes (A for baseline, B for SS, C for CA and D for DA). The classes have its own meaning: normal state (A), startle or surprise (SS), channelized attention (CA), and diverted attention (DA). With the classified mental states, pilots can be warned before falling into a dangerous state. By this way, we performed classification with XGBoot model along data clustering by k-means algorithm, that includes different k values (3,4,5,7 and 8). On the other hand, we also allowed learning by XGBoot for the standardized data without clustering for comparison.

D. Performance Report. Let us examine the experimental results exported by 5 folds k-means clustering and XGBoot classification. We can now explore the accuracy in each fold and its average value in Table 2, illustrated by Fig. 4. Note, as an interesting side issue, that the fifth fold with $k = 8$ has not provided result. We found that in the split, training set which created by the clustering task, contains

Table 2 Accuracy report for data in 5 folds

k	Fold 1	Fold 2	Fold 3	Fold 4	Fold 5	Average
No clustering	0.9404	0.9294	0.9360	0.9407	0.9278	*0.9349*
3	0.9439	0.9333	0.9419	0.9432	0.9321	0.9388
4	0.9342	0.9255	0.9383	0.9481	0.9438	0.9380
5	0.9324	0.9241	0.9423	0.9514	0.9408	0.9382
6	0.9400	0.9252	0.9387	0.9459	0.9451	0.9390
7	0.9383	0.9345	0.9496	0.9516	0.9363	**0.9421**
8	0.9304	0.9347	0.9472	0.9330	NA	0.9363

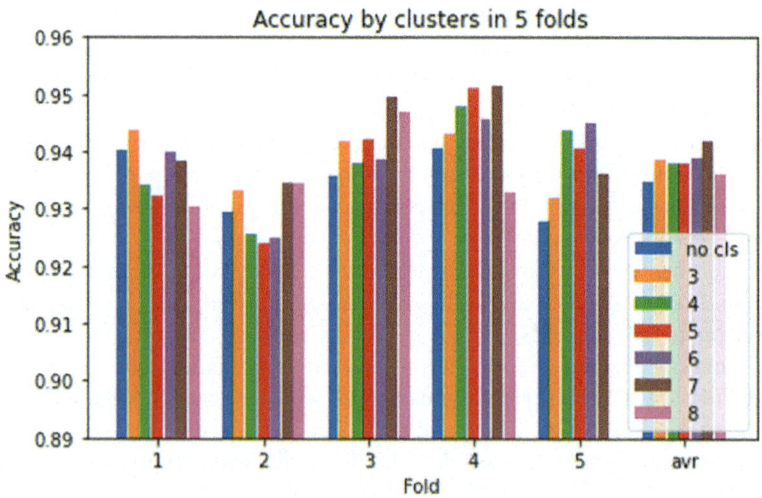

Fig. 4 Accuracy Report

samples from only one class. So the case was not appropriate for prediction and "not available" (na) is shown in the table.

Notice that, the classification in the absence of clustering has given averaged accuracy of 0.9349 which is printed in Italic. Actually the score is not bad but it is the smallest value in the column. As it happens, the case of clustering with 7 clusters yielded the highest accuracy by 0.9421 which is typed in bold in the table.

Now that we have obtained the three metrics using different number of clusters in the learning task, what do we do see? Table 2 displayed the final results in accuracy, precision and f1-score for each specific number of clusters. where the best scores are printed in bold. In Fig. 5, which illustrate the Table 3 by a color bars chart, the best precision score (0.9334) belongs to the case of four clusters. All other metrics have the highest values when using seven clusters and the scores of accuracy and f1-score are 0.9421 and 0.9355, accordingly.

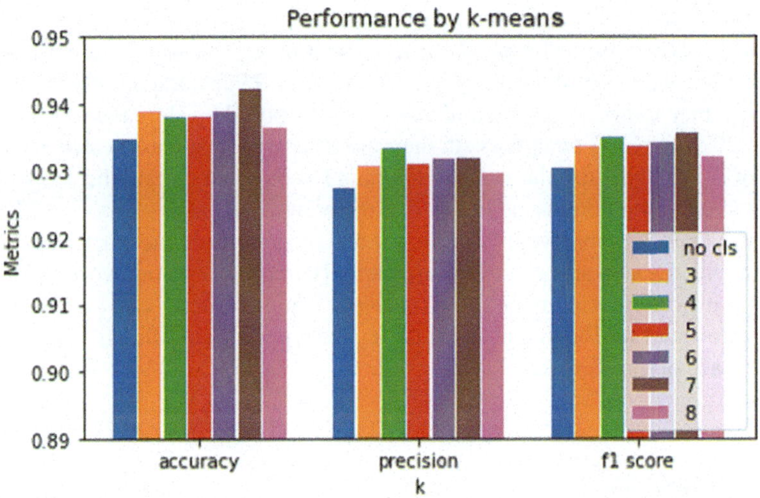

Fig. 5 Performance report

Table 3 Performance report

k	Accuracy	Precision	f1-score
No clustering	0.9349	0.9273	0.9304
3	0.9388	0.9308	0.9338
4	0.9380	**0.9334**	0.9352
5	0.9382	0.9309	0.9337
6	0.9390	0.9319	0.9342
7	**0.9421**	0.9319	**0.9355**
8	0.9363	0.9296	0.9320

*The best scores are printed in bold

Table 4 Other results

No	Predictor	Database	Accuracy	f1-score
1	Random Forest, 3 features [10]	Kaggle [9]	0.9082	0.9253
2	Random Forest, 4 features [10]	Kaggle [9]	0.9087	0.9254
3	Random Forest, all features [10]	Kaggle [9]	0.8976	0.9173
4	Neural network [12]	A320 flight simulator	0.8781	Na
5	Our model	Kaggle [9]	0.9421	0.9355

A specially resumed table was created for related works which used the same performance metrics (see Table 4). To predict mental states of pilots, Lin [10] used the dataset from the Kaggle [9] with assistance of Logistic Regression, Support Vector Machine, k-Nearest Neighbors, Recursive Partitioning and Regression Trees, Random Forests which given accuracy and f1-scores by 0.9087 and 0.9254 accordingly. More recently, convolutional neural network is proposed by Hernández-Sabaté et al. [12] in 2022 for EEG analysis with their own database that was created with assistance of A320 flight simulator. It is important to comment on that the shown scores are not suitable for comparing because of different databases.

However, as it is observed, the accuracy and other performance metrics for classification with the k-means clustering are partially improved. In our work, we showed results with different number (k) of clusters. The parameter can be chosen from experiments and we found that seven is the most appropriate value for our case with the database of the Kaggle [9].

5 Conclusions

Accuracy and complexity of the time series EEG data are of fundamental significance in quantitative pilot mental workload analysis, but there are relative challenge of such complex and big data. This article has demonstrated that our model can contribute for detecting metal states of crew members from physiological data. First, the memory used for storing data can be reduced. The representation of signals in restraint memory and noise removal allowed ameliorating computing speed. Second, reducing the dimension of data was yielded by constructive features extraction with PCA and autoencoder. Finally, the implementation of k-means clustering before classification by the extreme gradient boosting provided improvement in prediction performance.

References

1. C. Diaz-Piedra, M.V. Sebastián, L.L. Di Stasi, EEG theta power activity reflects workload among army combat drivers: an experimental study. Brain Sci. **28**;10(4), 199 (2020). https://doi.org/10.3390/brainsci10040199. PMID: 32231048; PMCID: PMC7226148
2. S. Lee, J.K. Kim, Factors contributing to the risk of airline pilot fatigue. J. Air Transp. Manage. **67**, 197–207 (2018). https://doi.org/10.1016/j.jairtraman.2017.12.009
3. D. Li, X. Wang, C.C. Menassa, V.R. Kamat, Understanding the impact of building thermal environments on occupants' comfort and mental workload demand through human physiological sensing. In Start-Up Creation; Elsevier: Amsterdam. The Netherlands **2020**, 291–341 (2020)
4. P. Zhang, X. Wang, W. Zhang, J. Chen, Learning spatial-spectral-temporal EEG features with recurrent 3D convolutional neural networks for cross-task mental workload assessment. IEEE Trans. Neural Syst. Rehabil. Eng. **27**, 31–42 (2019)
5. Ian T. Jolliff, Cadima Jorge, Principal component analysis: a review and recent developments. Philos. Trans. R. Soc. A: Math., Phys. Eng. Sci. **374**(2065), 20150202 (2016)
6. Z. Mengyue, W. SiuTim, L. Mengqi, Z. Tong, C. Kani, Time series generation with masked autoencoder (2022), arXiv:2201.07006
7. J. Crijnen, Predicting a pilot's cognitive state from physiological measurements. Master thesis, Tilburg University, Tilburg, The Netherlands (2019), http://arno.uvt.nl/show.cgi?fid=149399
8. A. Mishra, K.K. Shrivastava, A.B. Anto, N.A. Quadir, Reducing commercial aviation fatalities using support vector machines, in *2019 International Conference on Smart Systems and Inventive Technology (ICSSIT)* (IEEE, 2019), pp. 360–364
9. Kaggle, Reducing commercial aviation fatalities. Booz Allen Hamilton, Accessed April 22, 2020 (2019), https://www.kaggle.com/c/reducing-commercial-aviation-fatalities
10. Y.-C. Lin, Reducing aviation fatalities by monitoring pilots' cognitive states using psychophysiological measurements. Thesis, Naval Postgraduate School, Monterey, California (2021)
11. A.R. Harrivel, C.L. Stephens, R.J. Milletich, C.M. Heinich, M.C. Last, N.J. Napoli, N. Abraham, L.J. Prinzel, M.A. Motter, A.T. Pope, Prediction of cognitive states during flight simulation using multimodal psychophysiological sensing, in *AIAA Information Systems* (AIAA Infotech@Aerospace, 2017), https://doi.org/10.2514/6.2017-1135
12. A. Hernandez-Sabate, J. Yauri, P. Folch, M.A. Piera, D. Gil, Recognition of the mental workloads of pilots in the cockpit using EEG signals. Appl. Sci. (Switzerland) **12**(5), 2298 (2022). https://doi.org/10.3390/app12052298
13. B. Binias, D. Myszor, H. Palus, K.A. Cyran, Prediction of Pilot's reaction time based on EEG signals. Front Neuroinf. **14**(14), 6 (2020). https://doi.org/10.3389/fninf.2020.00006. PMID: 32116630; PMCID: PMC7033428
14. Sagi Omer, Rokach Lior, Approximating XGBoost with an interpretable decision tree. Inf. Sci. **572**(2021), 522–542 (2021). https://doi.org/10.1016/j.ins.2021.05.055
15. L.E. Ismail, W. Karwowski, Applications of EEG indices for the quantification of human cognitive performance: a systematic review and bibliometric analysis. PLoS One. **4**;15(12), e0242857 (2020). https://doi.org/10.1371/journal.pone.0242857. PMID: 33275632; PMCID: PMC7717519
16. M.L. Moroze, M.P. Snow, Causes and remedies of controlled flight into terrain in military and civil aviation. Technical report, Air Force Research Lab Wright-Patterson AFB OH, Human Effectiveness Directorate (1999)
17. W. Rosenkrans, Airplane state awareness. Flight safety foundation (2015), https://flightsafety.org/asw-article/airplane-state-awareness

Why Decreased Gaps Between Brain Cells Cause Severe Headaches: A Symmetry-Based Geometric Explanation

Laxman Bokati, Olga Kosheleva, Vladik Kreinovich, and Nguyen Hoang Phuong

Abstract When we analyze biological tissue under the microscope, cells are directly neighboring each other, with no gaps between them. However, a more detailed analysis shows that in vivo, there are small liquid-filled gaps between the cells, and these gaps are important: e.g., in abnormal situations, when the size of the gaps between brain cells decreases, this leads to severe headaches and other undesired effects. At present, there is no universally accepted explanation for this phenomenon. In this case, we show that the analysis of corresponding geometric symmetries leads to a natural explanation for this effect.

1 Decreased Gaps Between Brain Cells Cause Severe Headaches: Empirical Fact

Empirical fact. It is well known that living creatures consist of cells, cells is all you see when you look at any tissue under a microscope—and this is how scientists understood the structure of the living creatures. However, starting with the 1960s, it was determined that in living creatures, there is always a liquid-filled gap between

L. Bokati · V. Kreinovich (✉)
Computational Science Program, University of Texas at El Paso, 500 W. University, El Paso, TX 79968, USA
e-mail: vladik@utep.edu

L. Bokati
e-mail: lbokati@miners.utep.edu

O. Kosheleva
Department of Teacher Education, University of Texas at El Paso, 500 W. University, El Paso, TX 79968, USA
e-mail: olgak@utep.edu

N. Hoang Phuong
Artificial Intelligence Division, Information Technology Faculty, Thang Long University, Nghiem Xuan Yem Road, Hoang Mai District, Hanoi, Vietnam

N. H. Phuong and V. Kreinovich (eds.), *Deep Learning and Other Soft Computing Techniques*, Studies in Computational Intelligence 1097,
https://doi.org/10.1007/978-3-031-29447-1_3

cells—a gap that closes when the matter is no longer alive. The gaps were first discovered when, to preserve the structure as much as possible, researchers instantaneously froze the cells culture. Since then, new techniques have been developed to study these gaps.

The resulting studies showed that these gaps play important biological roles. For example, in abnormal situations, when the gaps between brain cells drastically decrease, a patient often experiences several headaches and other undesired effects; see [2] and references therein.

Why? As of now, there are no universally accepted explanations for this empirical fact. In this paper, we show that this empirical phenomenon naturally follows from symmetry-based geometric ideas.

2 Analysis of the Problem and the Resulting Explanation

Normal case: geometric description. Let us first consider the case when there are gaps between cells. We want to know how cells interact with each other.

All interactions are local. Thus, to analyze the interaction between the cells, we need to consider a small area on the border between two neighboring cells.

The border of each cell is usually smooth. Locally, each smooth surface can be well-approximated by its tangent plane—the smaller the area, the more accurate the approximation. Thus, with good accuracy, we can locally represent the border of each cell by a plane.

The border of the neighboring cell is also represented by a plane. Since we consider the situation when there is a gap, the borders do not intersect. When the two planes do not intersect, this means that they are parallel to each other. Thus, the normal configuration can be described by two parallel planes corresponding to two neighboring cells.

Abnormal case: geometric description. In the abnormal case, the gaps drastically decrease, to the extent these gaps become undetectable. Thus, with good accuracy, we can conclude that in this case, there is, in effect, no gaps between the two cells—and thus, that both cells can be described by a single plane, the plane that serves as a common border of the two cells.

We want to study dynamics. For a living creature, there is a usually a stable state, and then there are dynamic changes, when a change in one cell causes changes in others. To study dynamics, we therefore need to study how disturbances propagate. Since, according to physics, all interactions are local (see, e.g., [1, 3]), for a perturbation in one cell to reach another cell, this perturbation first need to reach the border of the original cell.

The simplest perturbation is when the perturbation is located at a single point on the cell's border—every other perturbation of the border can be viewed as a combination of such point-wise perturbations corresponding to all affected points.

So, to study how general perturbations propagate, it is necessary to study how point-wise perturbations propagate.

Role of symmetries. Most physical processes do not change if we apply:

- shift,
- rotation, or
- scaling—i.e., replace, for some $\lambda > 0$. each point with coordinates $x = (x_1, x_2, x_3)$ with a point with coordinates $(\lambda \cdot x_1, \lambda \cdot x_2, \lambda \cdot x_3)$.

Physical processes are invariant with respect to such geometric symmetries. Thus, if we start with the initial configuration which is invariant with respect to some of these symmetries, the resulting configuration will also be invariant with respect to the same geometric transformations. Let us analyze how this idea affects the propagation of perturbations between the cells.

Normal case: what are the symmetries and what are possible dynamics. Let us first consider the normal case, when we have:

- two parallel planes and
- a point in one of these planes—the location of the original perturbation.

One can see that out of all above-listed geometric symmetries, the only symmetries that keep this configuration invariant are rotations around the fixed point in the first plane—i.e., in 3D terms, rotations around the axis that goes through this point orthogonally to the plane (and, since the planes are parallel, orthogonally to both planes).

Since the initial configuration has this symmetry, the resulting configuration—observed after some time—should also has the same symmetry. In particular, with respect to what perturbations we can have on the other plane—i.e., on the border of the neighboring cell:

- we can have a single point (on the same axis),
- or we can have a rotation-invariant planar region (e.g., a disk centered at this point) that reflects inevitable diffusion.

These cases correspond to usual information transfer between the cells.

One of the possibilities is that the resulting configuration will involve all the points of the second plane, but this is not the only configuration resulting from diffusion.

Abnormal case: what are the symmetries and what are possible dynamics. Let us now consider the abnormal no-gaps case, when we have:

- a single plane—a joint boundary between the cells, and
- a point in this plane, which is the location of the original perturbation.

In this case, in addition to rotations, the corresponding configuration has an additional symmetry—scalings around the given point. Thus, the resulting configuration should be invariant not only with respect to the rotations, but also with respect to these scalings.

Due to inevitable diffusion, we expect the plane part of the resulting configuration to include more than a single point. However, one can easily see that every two points on a plane—which are both different from the original point—can be obtained from each other by an appropriate rotation and scaling. Thus, once the resulting rotation- and scale-invariant configuration contains at least one point which different from the original point, it will automatically include all the points in the plane.

In other words, in the presence of even small diffusion, a local point-wise perturbation will lead to a perturbation of the whole boundary. This perturbation will spread to other cells—and cause a global all-cells-involving perturbation, which is exactly what corresponds to a severe headache, when many cells are affected.

Summarizing: this explains the observed phenomenon. Our analysis shows the following.

- In the normal case—when there are gaps between brain cells—while we can have global-brain effects like severe headache, this is not inevitable: we can also have a usual information transfer between cells.
- On the other hand, in the no-gaps case, effects like severe headache are inevitable.

Thus, the thinner the gaps, the closer the resulting configuration is to the no-gaps case, the more probable it is that severe headaches (and other global effects) will occur—which is exactly what is observed. So, our symmetry-based geometric analysis indeed explains the observed phenomenon.

Acknowledgements This work was supported in part by the National Science Foundation grants 1623190 (A Model of Change for Preparing a New Generation for Professional Practice in Computer Science), and HRD-1834620 and HRD-2034030 (CAHSI Includes), and by the AT&T Fellowship in Information Technology.
It was also supported by the program of the development of the Scientific-Educational Mathematical Center of Volga Federal District No. 075-02-2020-1478, and by a grant from the Hungarian National Research, Development and Innovation Office (NRDI).

References

1. R. Feynman, R. Leighton, M. Sands, *The Feynman Lectures on Physics* (Addison Wesley, Boston, 2005)
2. C. Nicholson, The secret world in the gaps between brain cells. Phys. Today **75**(5), 26–32 (2022)
3. K.S. Thorne, R.D. Blandford, *Modern Classical Physics: Optics, Fluids, Plasmas, Elasticity, Relativity, and Statistical Physics* (Princeton University Press, Princeton, 2017)

Drug Repositioning for Drug Disease Association in Meta-paths

Xuan Tho Dang, Manh Hung Le, and Nam Anh Dao

Abstract The identification of potential interactions and relationships between drugs and diseases gains strength in recognition in the field of biomedical signal processing. When experiments for determining the drug and disease association usually require intensive investment and time, the new method of drug repositioning could assist in cutting-off cost and minimizing risk of failure. In contrast to the conventional implementation, the data in this problem are heterogeneous due to integration of different data sources. We propose to apply a robust meta-path for dealing with the heterogeneous data in the problem of drug repositioning. A set of five new meta paths based on the heterogeneous biological network of drug-protein-disease is implemented to define new drug-disease association. Technical advances in implementing the new meta paths are clear logical structure of the meta-paths and improvement in several performance metrics.

Keywords Drug-disease associations · Knowledge graph embedding · Structural representation · Drug repositioning

1 Introduction

The targets of drug discovery to identify potential new medicines for a specific disease. This can be performed by involving a wide range of research domains including biology, chemistry pharmacology and computing science.

Actually, the cost for drug discovery is usually high, especially when the process is laborious, time-consuming and highly risk process. It is possible to enhance the success rate of the drug discovery by drug repositioning which examines existing

X. Tho Dang
Hanoi National University of Education, Hanoi, Vietnam

M. Hung Le · N. Anh Dao (✉)
Electric Power University, Hanoi, Vietnam
e-mail: anhdn@epu.edu.vn

N. H. Phuong and V. Kreinovich (eds.), *Deep Learning and Other Soft Computing Techniques*, Studies in Computational Intelligence 1097,
https://doi.org/10.1007/978-3-031-29447-1_4

drugs for novel therapeutic targets. There was accident finding a novel induction in drug history. For instance, Minoxidil, which was first used to treat hypertension, was unintentionally discovered to be effective in treating hair loss [1]; Sildenafil (brand name: Viagra), which was first used to treat angina, was occasionally discovered to have the ability to treat erectile dysfunction [2]. However, the "pot-luck" strategy cannot guarantee that drugs will be re-positioned effectively and successfully.

Traditional medication research and development is a drawn-out procedure that incurs significant expenses and, very few medications reach the market, even after receiving approval [3]. Pharmaceutical experts have recently concentrated heavily on cutting-edge drug discovery methods that draw on information of already-approved medications. The drug repositioning is considered as a capable method to develop drug candidates with new therapeutic properties or pharmacological activities. In particular, the current approved and clinically used drug molecules are investigated for question if they interact with protein of a specific disease.

Drug repositioning, also known as drug re-purposing, is a promising method for identifying new markers for currently used or experimental medications, and it has the potential to significantly speed up the development of novel drugs [4]. In this work, we concentrate specifically on drug-disease association (DDA) which comes up with significant information for discovering the potential efficacy of drugs. Here, to establish novel potential DDA, it is essential to apply learning algorithms for analyzing relationships between diseases and drugs. The machine learning based approach has advantages over traditional drug discovery method allowing to decrease duration of drug development, to cut-off cost and minimize risk of failure.

Numerous computational-based methods for finding fresh indicators of approved drugs have been developed recently [5–9] for getting data source for drug repositioning. By predicting potential treatment linkages of medication-disease combinations, these approaches can generally be divided into three categories: network-based [10, 11], matrix factorization-based [12–14], and deep learning-based approaches [15, 16].

Our method is based on the same assumptions as in [14], where non-associated drug and disease pair keep no shared interconnected proteins, which is distinct from other methods. As Wu et al. noted in [14], specifying more likely negative ones from unlabeled samples for training could yield a more accurate forecast model.

In this work, we propose new five meta paths based on the heterogeneous biological network of drug-protein-disease. For each meta path, we implement a machine learning model, then an integrated learning method is formed by the models. Our first finding is that each distinct meta path, which checks drug—protein—disease relation in a specific way is capable to detect what relation can be potential for repositioning drug. Second, combining several meta paths on an heterogeneous data in a particular logical way, validity of repositioning drugs can be improved. We evaluated our approach on the a data set extracted from different resources. It has been demonstrated that the experiments in this paper can achieve advanced performance.

2 Related Work

It is important to understand what data the repositioning drug currently considers and which approaches recent research reports used for repositioning drug. The section involves two parts: feature similarity and network-based methods.

A. Feature similarity. On the basis of machine learning techniques, numerous methods have been presented to integrate various data sources in order to predict the drug-disease interactions connected to multi-omics data. With regard to illness treatment, symptomatology, or pathology, the disease-centric approach primarily makes use of the characteristics of diseases [17]. This method either detects and uses the common traits of diseases connected with an existing medicine or it develops a group of disorders, by merging current knowledge about diseases.

Gottlieb et al. indirectly employed the characteristics of medications and disorders as another drug-disease mutual approach [18]. They created drug-drug and disease-disease similarity measurements based on the notion that comparable medications are indicated for similar disorders, then used these measures to create classification features and develop a classification rule. It is only possible for reproducible implementation if all necessary pharmacological and disease attributes are gathered. Functional genomic networks are becoming more precise and comprehensive with the growing number and variety of high-throughput datasets. By integrating heterogeneous data, Martinez et al. created the DrugNet technique for drug-disease and disease-drug prioritization [19].

To forecast the relationship between medications and diseases, Ying Wang et al. used neighborhood information aggregation in neural networks and integrates the connections between diseases and drugs as well as the similarity of diseases and drugs [20]. Three matrices-the drug similarity matrix, the disease similarity matrix, and the drug and disease association matrix-were combined into a single, sizable matrix by H. Luo et al. Then it determines the big matrix's lowest level, from which the large matrix DRRS is rebuilt [21].

NeoDTI, which was proposed by F. Wan et al. is based on end-to-end learning using a nonlinear model, predicts new medications and drug targets by integrating varied inputs in heterogeneous networks [22]. In order to infer novel treatments for diseases using a heterogeneous network incorporating similarity and association data about diseases, drugs, and therapeutic targets, W. Wang et al. introduced the computational framework TL-HGBI [23]. Using vertex-clustering preferences and the likelihood of feature-clustering contributions, MrSBM also takes into account modeling vertex features in addition to modeling edges that are contained within or connect blocks [24].

A DDA prediction approach based on matrix factorization called SCMFDD sets limitations on illness semantic similarity and drug similarity increase while mapping drug-disease relationships into low-rank space [25]. Only medications and disease information were chosen for binary network, which Zhang et al. utilized to predict DDAs [26]. Drug repositioning computation is being steadily solved from a macro perspective, however past studies on DDA prediction did not take the entire cell into account.

In most problems of feature similarity is the requirement of information about how similar drugs, proteins and diseases. This is not always available. Our approach of using meta path permits extract DDA feature without the need of searching the feature similarity.

B. Network-based approaches. In the field of drug repositioning, there have been discussed many clustering algorithms with many existing computational methods need to compute the similarities of drugs and diseases. They mainly aim can mine out many drug-disease associations. In order to merge various drug properties and disease features, Moghadam et al. used the kernel fusion technique [27]. After that, they created SVM models to predict novel pharmacological indications. He et al. proposed an algorithm, which takes into account the topology structure and attribute value of the graph, can successfully identify more relevant clustering hidden in the attribute graph [28].

A computation approach for DDA predictions has been proposed by Hanjing Jiang at al. basing on graph representation learning over multi-biomolecular network (GRLMN) [29]. He et al. showed MrSBM [24] for working with network data to conduct unsupervised learning tasks. To predict the drug-disease connections, Wu et al. included complete drug-drug and disease-disease similarities from chemical/phenotype layer, gene layer, and treatment network layer [30]. They presented a semi-supervised graph cut approach (SSGC).

NeoDTI, which was proposed by F. Wan et al., is based on end-to-end learning using a nonlinear model, for predicting new medications and drug targets by integrating varied inputs in heterogeneous networks. In order to infer novel treatments for diseases using a heterogeneous network incorporating similarity and association data about diseases, drugs, and therapeutic targets, W. Wang et al. introduced the computational framework TL-HGBI [22].

The relationship between possible medications and diseases was computed by A. Gottlieb et al. primarily by combining the similarities between various pharmaceuticals and disorders and applying these features to obtain new potential features through a logical classifier [18].

He et al. showed CCPMVFGC which combines graph clustering with multi-view learning for capturing the contextual interdependency of features in each cluster [31]. Layer Attention Graph Convolutional Network (LAGCN) is presented by Yu et al. to predict DDA. It uses graph convolution to learn DDA, drug-drug similarity, and disease-disease similarity, and uses the attention mechanism to merge multiple graph convolution layers [32].

It will appear that the meta paths in our solution using the same learning prediction model for five meta path and combine them in other five meta paths for getting performance enhanced.

3 The Method

In this section we describe the framework for predicting drug-disease associations and it contains four steps. Then, a heterogeneous network is build in the same way of [14] from three available biological databases: OMIM [33], Gottlieb [34] and DrugBank [35] datasets. Second, a feature vector was constructed based on each meta-path through above network.

From cross validation approach, we divided the drug-disease dataset into two parts: one for training, other for testing. In training, the our meta-paths were applied for the training dataset giving output with the same size of the original training dataset. These operations are displayed in the part of Step 1 in the Fig. 1.

There are drug-disease interactions registered by mentioned above biological databases and represented by an array A_{DS} where rows are indexed by drugs (D) and columns are indexed by diseases (S). With drug-protein interactions we use array A_{DP} to show registered interactions between drugs and proteins (P). Also in showing disease-protein interaction it is array A_{DP} displaying registered interactions. In fact, the three arrays A_{DS}, A_{DP}, A_{SP} are appeared in multiplication for described five meta paths in [14], and they are shown in the left of the part of Step 2 of the Fig. 1, allowing to have five output arrays $X1, X2, X3, X4, X5$.

$$X1 = A_{DS} \tag{1}$$

$$X2 = A_{DP} * A_{SP}^T \tag{2}$$

$$X3 = A_{DP} * A_{DP}^T * A_{DS} \tag{3}$$

$$X4 = A_{DS} * A_{DS}^T * A_{DS} \tag{4}$$

$$X5 = A_{DS} * A_{SP} * A_{SP}^T \tag{5}$$

Now, the arrays are used as input for new five meta paths which are proposed to be improved compared to the previous method to reflect different aspects of the drug-disease treatment relationship in heterogeneous networks. The task is shown in the center of the part of Step 2 of the Fig. 1 producing five output arrays $X1a, X2a, X3a, X4a, X5a$ which serves the Step 3 as input.

$$X1a = X1 * X2^T * X5 \tag{6}$$

$$X2a = X2 * X3^T * X5 \tag{7}$$

$$X3a = X3 * X4^T * X5 \tag{8}$$

$$X4a = X4 * X5^T * X5 \tag{9}$$

$$X5a = X5 * X1^T * X5 \tag{10}$$

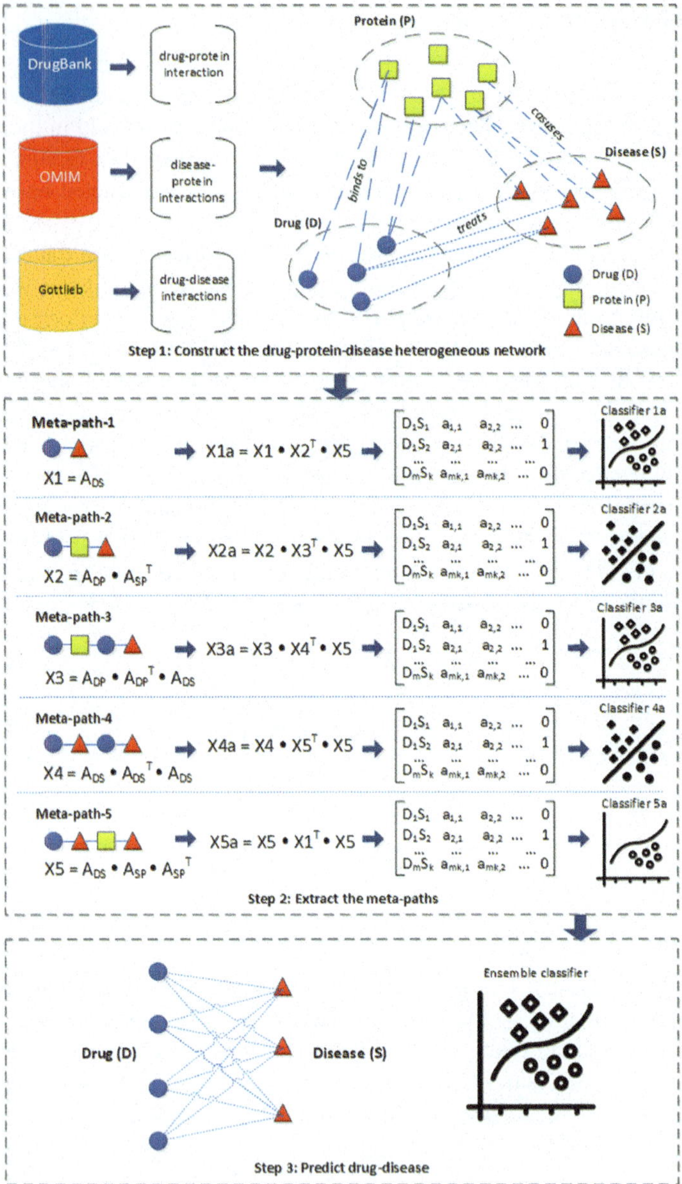

Fig. 1 Workflow of meta paths

Table 1 Data structure in experiments

Relation	Matrix shape	Number of interactions	Ratio in %
Drug-disease	(1186, 449)	1827	0.3431
Drug-protein	(1186, 1467)	4642	0.2668
Disease-protein	(449, 1467)	1365	0.2072

From these five feature array data, the Random Forest classifiers were constructed to predict drug-disease treatment relationships with different aspects. Finally, we proceed to integrate these classifiers into an ensemble classifier. The ensemble classifier calculates the positive probability for each drug-disease pair is related or not (negative/positive). This is illustrated by Step 3 in the Fig. 1.

Within the above context, the experiments of repositioning drugs in meta paths provides an assessment of the described method providing implementation on an actual database. The problem of finding a new possible interaction of drugs, proteins and diseases is equivalent to picking out appropriate paths of the original interaction matrices. Three interaction pairs of with binary values, correspond to three components domain: drug—protein, protein—disease and drug—disease. The data for our experiments, were combined by [14] where interaction data are extracted from OMIM [33], Gottlieb [34] and DrugBank [35].

Actually, the data have 1186 drugs, 449 diseases and 1467 proteins. However, only 1827 drug disease interactions were confirmed giving density of 0.34% for the matrix size (1186, 449). By checking that the drug—protein interactions are 4642, the density of known interactions is 0.26%, while the density of disease—protein is 0.20% by holding 1365 interactions (Table 1).

We further operate the cross validation by performing five folders of each meta path and the ensemble path. A number of classification metrics has been used to enable the performance evaluation, including the accuracy, which is described as the number of proper predictions divided by the total number of predictions, multiplied by 100:

$$Accuracy = \frac{n_{TP} + n_{TN}}{n_{TP} + n_{FP} + n_{TP} + n_{FN}}, \tag{11}$$

where TP, TN, FP and FN are the number of true positive samples, true negative samples, false positive samples and false negative samples, correspondingly. To give another view of performance, here the Precision (PRE), Recall (REC), Matthews Correlation Coefficient (MCC) and F1-measure (F1) were measured for each test split:

$$REC = \frac{n_{TP}}{n_{TP} + n_{FN}}, \tag{12}$$

$$REC = \frac{n_{TN}}{n_{TN} + n_{FP}}, \tag{13}$$

$$PRE = \frac{n_{TP}}{n_{TP} + n_{FP}}, \tag{14}$$

$$F1 = \frac{2 \times Recall * Precision}{Recall + Precision}, \tag{15}$$

$$MCC = \frac{n_{TP} * n_{TN} - n_{FP} * n_{FN}}{\sqrt{(n_{TP} + n_{FP})(n_{TP} + n_{FN})(n_{TN} + n_{FP})(n_{TN} + n_{FN})}}. \tag{16}$$

In order to compare the meta-paths in the term of true positive rate and the false positive rate, the Area Under Precision-Recall Curve (AUPR) and Area Under Receiver Operating Characteristic Curve (AUC) are also used in our experiments.

In particular, the described above metrics for our experiments are briefly summarized in Table 2. The first 5 lines show results for each meta path and the sixth row displays of the final ensemble path.

Next, the results of the new five meta paths are reported in the first 5 lines in the Table 3 allowing to have the final result in the sixth line. As can be observed, the results are ameliorated on the previous version.

These results highlight that all new five meta paths under review can be carried out to obtain the enhanced performance metrics in terms of adding meta-paths. The AUC are shown in Fig. 2, where five splits of the ensemble path are printed in different colors with area above 95%. This confirms that the new meta paths is strong enough to allow the drug repositioning processed.

Table 2 Results of meta paths from [14] by experiments

Path	AUPR	AUC	PRE	REC	ACC	MCC	F1
1	0.895	0.860	0.773	0.886	0.835	0.675	0.825
2	0.936	0.928	0.850	0.873	0.861	0.722	0.860
3	0.926	0.905	0.832	0.883	0.858	0.719	0.855
4	0.895	0.860	0.784	0.875	0.834	0.673	0.827
5	0.921	0.895	0.804	0.920	0.866	0.737	0.858
6	0.956	0.951	0.912	0.855	0.876	0.754	0.882

Table 3 Results of new meta paths by our experiments

Path	AUPR	AUC	PRE	REC	ACC	MCC	F1
1	0.927	0.903	0.835	0.886	0.862	0.724	0.860
2	0.945	0.942	0.899	0.874	0.883	0.766	0.886
3	0.957	0.950	0.887	0.907	0.895	0.790	0.896
4	0.924	0.898	0.824	0.907	0.867	0.738	0.863
5	0.921	0.896	0.809	0.917	0.866	0.737	0.859
6	0.962	0.956	0.882	0.901	0.889	0.780	0.889

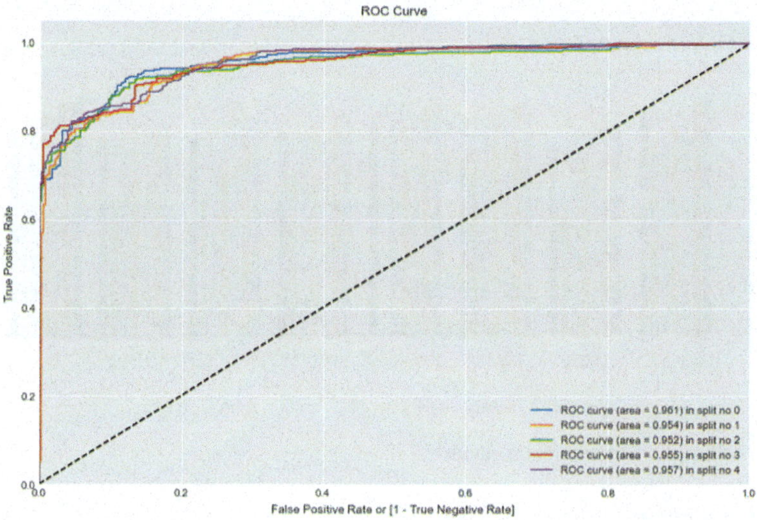

Fig. 2 AUC by the new meta path

As expected, the results of related works are resumed for making a comparative report. In the following the works are briefly described. To predict drug-disease relationships, Liang et al. combined drug chemistry information with target domain and gene ontology annotation information and suggested the Laplacian regularized sparse subspace learning approach (LRSSL) [36]. By otherhand, Luo et al. used a comprehensive comparison of medications and disorders and suggested the Bi-Random walk algorithm (MBiRW) [37] for predicting probable drug-disease interactions.

For predicting DDA Kawichai et al. [38] applied an ensemble learning method which processed meta-path. Gene ontology (GO) terms were generated for linking drugs and diseases to their associated functions. Hence the Go can be used as intermediate nodes in tripartite network. In order to predict drug-disease associations, Zhang et al. first introduced a linear neighborhood similarity [40] and a network topological similarity [39], then they proposed a similarity constrained matrix factorization method (SCMFDD) using information on known drug-disease associations, drug features, and disease semantics [25]. By including similarities and interactions among diseases, medications, and therapeutic targets, Wang et al. established a computational framework based on a three-layer heterogeneous network model (TL-HGBI) [41] (Fig. 3 and Table 4).

The approaches, currently in use, rely on the idea that diseases are treated by similar treatments, therefore they require information about how similar pharmaceuticals, proteins, diseases, and other things are. The similarity data, however, is not readily available [13]. To meet their individual objectives, people frequently need to modify a program that collects data and calculates similarities. The similarity data, however, is not readily available. To meet their individual objectives, people frequently need to modify a program that collects data and calculates similarities. A pair's similarity

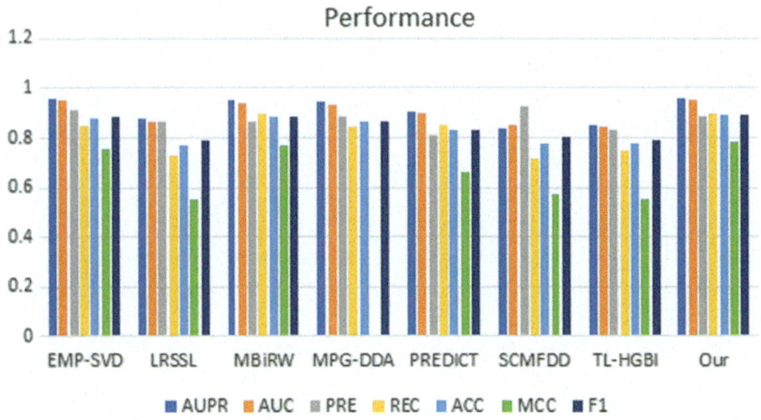

Fig. 3 Performance of different methods

Table 4 Performance of related methods

Methods	AUPR	AUC	PRE	REC	ACC	MCC	F1
EMP-SVD [14]	0.956	0.951	**0.913**	0.854	0.876	0.755	0.882
LRSSL [36]	0.881	0.861	0.864	0.732	0.770	0.553	0.79
MBiRW [37]	0.952	0.942	0.867	0.901	0.884	0.769	0.884
MPG-DDA [38]	0.944	0.93	0.886	0.842	0.867	NA	0.863
PREDICT [11]	0.908	0.895	0.809	0.85	0.830	0.662	0.828
SCMFDD [25]	0.836	0.854	0.926	0.713	0.774	0.575	0.805
TL-HGBI [41]	0.852	0.846	0.829	0.75	0.774	0.552	0.787
Our method	**0.962**	**0.956**	0.882	**0.901**	**0.889**	**0.780**	**0.889**

*The best scores are printed in bold

score may vary significantly between methods because the calculation of similarity scores depends on the metrics that have been used. For instance, two proteins may be distinct in terms of their sequences but comparable in terms of their shapes. Even worse, some features that are necessary for computing the similarities may not exist or may not be available, which makes these methods ineffective.

The other issue is that drug-disease pair data, like other biological data, lacks experimentally verified negative samples, whereas supervised learning-based approaches typically require both positive and negative samples to train the prediction models. Most of the currently used methods randomly choose some unlabeled data to serve as the negative samples while training the models. Unlabeled data may contain some positive samples, but this approach is obviously very shaky because we are unsure of this.

Using data on drug-disease, drug-protein, and disease-protein interactions, we provide a strategy in this work. We presented new five routes of meta paths to combine various types of interaction data and take into account various dependencies.

4 Conclusions

A new meta-path method has been developed for detecting new drug—disease association from existing association. It is based on the heterogeneous biological network of drug-protein-disease. No additional data is required, and our meta paths model is tolerant to within heterogeneous network, where the number of confirmed association is very small in comparison to the size of the network.

We have made contributions with our evaluation of meta paths and adding new five meta-paths based on the evaluation. Experimental results show that the new meta-paths presented here have the ability for determining potential drugs for disease. In particular, the work is robust with respect to heterogeneity of data.

Acknowledgements This research was supported by the Vietnam Ministry of Education and Training, project B2022-SPH-04.

References

1. S. Varothai, W.F. Bergfeld, Androgenetic alopecia: an evidence-based treatment update. Am. J. Clin. Dermatol. **15**(3), 217–30 (2014)
2. N. Novac, Challenges and opportunities of drug repositioning. Trends Pharmacol. Sci. **34**(5), 267–72 (2013)
3. S. Pushpakom, F. Iorio, P.A. Eyers et al., Drug repurposing: progress, challenges and recommendations. Nat. Rev. Drug Discovery **18**, 41–58 (2019)
4. Z. Liu, H. Fang, K. Reagan et al., In silico drug repositioning-what we need to know. Drug Discovery Today **18**, 110–115 (2013)
5. M. Bagherian, E. Sabeti, K. Wang et al., Machine learning approaches and databases for prediction of drug-target interaction: a survey paper. Brief. Bioinform. **22**, 247–269 (2021)
6. X. Su, L. Hu, Z. You et al., A deep learning method for repurposing antiviral drugs against new viruses via multi-view nonnegative matrix factorization and its application to SARS-CoV-2. Briefings Bioinform. **23**, bbab526 (2022)
7. X. Su, L. Hu, Z. You et al., Attention-based knowledge graph representation learning for predicting drug-drug interactions. Briefings Bioinform. **23**, bbac140 (2022)
8. P. Hu, Y.-A. Huang, J. Mei et al., Learning from low-rank multimodal representations for predicting disease-drug associations. BMC Med. Inform. Decis. Mak. **21**, 1–13 (2021)
9. L. Hu, J. Zhang, X. Pan et al., HiSCF: leveraging higher-order structures for clustering analysis in biological networks. Bioinformatics **37**, 542–550 (2021)
10. X. Wang, B. Xin, W. Tan et al., DeepR2cov: deep representation learning on heterogeneous drug networks to discover anti-inflammatory agents for COVID-19. Briefings Bioinform. **22**, bbab226 (2021)
11. X. Zeng, S. Zhu, X. Liu et al., deepDR: a network-based deep learning approach to in silico drug repositioning. Bioinformatics **35**, 5191–5198 (2019)
12. M. Yang, G. Wu, Q. Zhao et al., Computational drug repositioning based on multi-similarities bilinear matrix factorization. Briefings Bioinform. **22**, bbaa267 (2021)
13. Y. Luo, X. Zhao, J. Zhou et al., A network integration approach for drug-target interaction prediction and computational drug repositioning from heterogeneous information. Nat. Commun. **8**, 1–13 (2017)
14. Wu G., Liu J., and Yue X.(2019) Prediction of drug-disease associations based on ensemble meta paths and singular value decomposition, From The 17th Asia Pacific Bioinformatics Conference

15. X. Su, Z. You, L. Wang et al., SANE: a sequence combined attentive network embedding model for COVID-19 drug repositioning. Appl. Soft Comput. **111**, 107831 (2021)
16. L. Cai, C. Lu, J. Xu et al., Drug repositioning based on the heterogeneous information fusion graph convolutional network. Briefings Bioinform. **22**, bbab319 (2021)
17. J.T. Dudley, T. Deshpande, A.J. Butte, Exploiting drug-disease relationships for computational drug repositioning. Brief. Bioinform. **12**(4), 303–311 (2011)
18. A. Gottlieb, G.Y. Stein, E. Ruppin, R. Sharan, PREDICT: a method for inferring novel drug indications with application to personalized medicine. Mol. Syst. Biol. **7**(1), 496 (2014)
19. V. Martínez, C. Navarro, C. Cano et al., (2015) Drugnet: Network-based drug-disease prioritization by integrating heterogeneous data. Artif. Intell. Med. **63**(1), 41–9 (2015)
20. Y. Wang et al., Drug-disease association prediction based on neighborhood information aggregation in neural networks 50581–50587 (2019)
21. H. Luo, M. Li, S. Wang, Q. Liu, Y. Li, J. Wang, Computational drug repositioning using low-rank matrix approximation and randomized algorithms. Bioinformatics **34**(11), 1904–1912 (2018)
22. F. Wan, L. Hong, A. Xiao, T. Jiang, J. Zeng, NeoDTI: Neural integration of neighbor information from a heterogeneous network for discovering new drug-target interactions. Bioinformatics **35**(1), 104–111 (2018)
23. W. Wang, S. Yang, X. Zhang, J. Li, Drug repositioning by integrating target information through a heterogeneous network model. Bioinformatics **30**(20), 2923–2930 (2014)
24. T. He, L. Bai, Y.-S. Ong, Manifold regularized stochastic block model, in *2019 IEEE 31st International Conference on Tools with Artifcial Intelligence (ICTAI)* (IEEE, 2019), pp. 800–807
25. W. Zhang, X. Yue, W. Lin, W. Wu, R. Liu, F. Huang, F. Liu, Predicting drug-disease associations by using similarity constrained matrix factorization. BMC Bioinform. **19**(1), 233 (2018)
26. W. Zhang, X. Yue, Y. Chen, W. Lin, B. Li, F. Liu, X. Li, Predicting drug-disease associations based on the known association bipartite network, in *2017 IEEE International Conference on Bioinformatics and Biomedicine (BIBM)* (IEEE, 2017), pp. 503–509
27. H. Moghadam, M. Rahgozar, S. Gharaghani, Scoring multiple features to predict drug disease associations using information fusion and aggregation. SAR QSAR Environ. Res. **27**(8), 609–28 (2016)
28. T. He, K.C. Chan, Discovering fuzzy structural patterns for graph analytics. IEEE Trans. Fuzzy Syst. **26**(5), 2785–96 (2018)
29. H. Jiang, Y. Huang, An effective drug-disease associations prediction model based on graphic representation learning over multi-biomolecular network. BMC Bioinform. (2022)
30. G. Wu, J. Liu, C. Wang, Predicting drug-disease interactions by semi-supervised graph cut algorithm and three-layer data integration. BMC Med. Genet. **10**(5), 79 (2017)
31. T. He, Y. Liu, T.H. Ko, K.C.C. Chan, Y. Ong, Contextual correlation preserving multiview featured graph clustering. IEEE Trans. Syst. Man Cybern. 1–14 (2019)
32. Z. Yu, F. Huang, X. Zhao, W. Xiao, W. Zhang, Predicting drug-disease associations through layer attention graph convolutional network. Brief Bioinform. **22**(4), 243bbaa (2020)
33. N.M. O'Boyle, M. Banck, C.A. James et al., Open babel: an open chemical toolbox. J Cheminform. **3**(1), 33 (2011)
34. T.F. Smith, M.S. Waterman, C. Burks, The statistical distribution of nucleic acid similarities. Nucleic Acids Res. **13**(2), 645–56 (1985)
35. W.D. Smiles (1988) a chemical language and information system. 1. introduction to methodology and encoding rules. J. Chem. Inf. Comput. Sci. **28**(1), 31-36
36. X. Liang, P. Zhang, L. Yan et al., Lrssl: predict and interpret drug-disease associations based on data integration using sparse subspace learning. Bioinformatics (Oxford, England). **33**(8), 1187–1196 (2017)
37. H. Luo, J. Wang, M. Li et al., Drug repositioning based on comprehensive similarity measures and bi-random walk algorithm. Bioinformatics **32**(17), 2664–2671 (2016)
38. T. Kawichai, A. Suratanee, K. Plaimas, Meta-path based gene ontology profiles for predicting drug-disease associations. IEEE Access **9**, 41809–41820 (2021). https://doi.org/10.1109/ACCESS.2021.3065280

39. W. Zhang, X. Yue, F. Huang et al., Predicting drug-disease associations and their therapeutic function based on the drug-disease association bipartite network. Methods **145**, 51–59 (2018)
40. W. Zhang, X. Yue, F. Liu et al., A unified frame of predicting side effects of drugs by using linear neighborhood similarity. BMC Syst. Biol. **11**(6), 101 (2017)
41. W. Wang, S. Yang, X. Zhang et al., Drug repositioning by integrating target information through a heterogeneous network model. Bioinformatics **30**(20), 2923–30 (2014)

iR1mA-LSTM: Identifying N^1-Methyladenosine Sites in Human Transcriptomes Using Attention-Based Bidirectional Long Short-Term Memory

Trang T. T. Do, Thanh-Hoang Nguyen-Vo, Quang H. Trinh, Phuong-Uyen Nguyen-Hoang, Loc Nguyen, and Binh P. Nguyen

Abstract Methylation is the most frequently occurring epigenetic modification that accounts for over 50% of the total modification forms. Among the methylation sites in the adenosine nucleobase, N1-methyladenosine (1mA) is a significant post-transcriptional alteration found in myriad types of RNA molecules. The alteration of gene sequences caused by this methylation affects numerous biological processes, such as genome editing, cellular differentiation, and gene expression, causing dangerous diseases. Many experimental techniques were developed to determine 1mA in RNA sequences. These approaches, however, are not cost- and time-effective solutions for wide screening in laboratories with limited budgets. To partially address this problem, several computational methods have been introduced to assist experimental scientists in identifying 1mA sites. In this paper, we present a more effective computational model called iR1mA-LSTM to predict 1mA sites in human transcriptomes using the Long Short-term Memory networks enhanced by an attention mechanism

---

T. T. T. Do
School of Innovation, Design and Technology, Wellington Institute of Technology, Lower Hutt 5012, New Zealand
e-mail: trang.do@weltec.ac.nz

T.-H. Nguyen-Vo · L. Nguyen · B. P. Nguyen (✉)
School of Mathematics and Statistics, Victoria University of Wellington, Wellington 6140, New Zealand
e-mail: binh.p.nguyen@vuw.ac.nz

T.-H. Nguyen-Vo
e-mail: thanhhoang.nguyenvo@vuw.ac.nz

L. Nguyen
e-mail: khanhloc.nguyen@vuw.ac.nz

Q. H. Trinh
School of Information and Communication Technology, Hanoi University of Science and Technology, 100000 Hanoi, Vietnam

P.-U. Nguyen-Hoang
Computational Biology Center, International University, VNU HCMC, Ho Chi Minh City 700000, Vietnam

N. H. Phuong and V. Kreinovich (eds.), *Deep Learning and Other Soft Computing Techniques*, Studies in Computational Intelligence 1097,
https://doi.org/10.1007/978-3-031-29447-1_5

to improve the predictive power. We conducted repeated experiments to fairly estimate the robustness and stability of the model performance and benchmarked our model with other methods on the same independent test set. The results show that iR1mA-LSTM outperformed other methods with both the area under the receiver operating characteristic curve and the area under the precision-recall curve values of over 0.99. Links to our source code and a web server implementing iR1mA-LSTM are available at https://github.com/mldlproject/2022-iR1mA-LSTM.

1 Introduction

The ribonucleic acid (RNA) family has a wide range of functions in living cells. Although RNA modifications frequently occur, understanding their mechanisms and biological roles remains superficial [1]. Among over 100 types of known RNA modifications, methylation is the most widespread form, accounting for over 50% of the total forms [2]. There are four methylation sites in the adenosine nucleobase, one of which, N1-methyladenosine (1mA) is a significant post-transcriptional alteration found in diverse types of RNA molecules, especially tRNA and rRNA [3]. N1-methyladenosine was found to involve in a large number of disorders and pathogenesis processes [3]. This methylation type hampers RNA metabolism and activities, including RNA synthesis, structural maintenance, stabilizing, and translation [4]. When 1mA is out of control, it may mislead gene expression to abnormalize physiological activities [4]. Therefore, attaining holistic insights about 1mA is crucial to discovering its unexplored mechanisms and biological functions.

To detect 1mA sites in RNA sequences, multiple experimental approaches have been developed based on their specific characteristics. Adenines at 1mA sites exhibit positive charges under physiological conditions, and this response may affect secondary structures of RNAs or RNA-protein interactions. For understanding their molecular behaviors, high-throughput sequencing techniques, such as bi-dimensional thin-layer chromatography or high-pressure liquid chromatography, were developed to identify possible 1mA alteration identification [5]. To improve sensitivity in quantifying the 1mA level in RNAs, liquid chromatography and mass spectrometry were incorporated into one platform. On the other hand, since the 1mA-nucleosides are also able to create hydrogen bonds with other nucleosides through a base pairing process called Hoogsteen, the 1mA-nucleosides can give rise to misincorporation or truncation during reverse transcription (RT) reaction [6, 7]. The 1mA-nucleoside' signature in RT reaction was used to design RT-qPCR-based assays (e.g., RTL-P) to identify and pinpoint the alteration [6, 7]. Although these techniques are no longer unfamiliar to many laboratories, their cost is still a financial concern for low-budget laboratories in developing countries when applying wide-screening. Additionally, to effortlessly use these techniques and assure the output quality, experienced staff are required.

To partially address this issue, multiple computational methods have been proposed to assist experimental scientists in identifying 1mA sites in RNA sequences [8–

10]. RAMPred [8] was proposed to identify the 1mA sites in the transcriptomes of *Homo sapiens* (*H. sapiens*), *Mus musculus* (*M. musculus*) and *Saccharomyces cerevisiae* (*S. cerevisiae*). This model was developed using support vector machines (SVM) and nucleotide chemical property and composition features. Another SVM-based model, iRNA-3typeA [9], was also developed to predict multiple types of RNA modification at Adenosine sites. Most recently, ISGm1A [10] was introduced to solve this biological issue using the random forest algorithm incorporated with sequence and genomic features for sequence representation. Cross-validation was applied in all three methods while testing using independent data was performed by ISGm1A only. These computational approaches share common features in employing classical machine learning algorithms. However, classical machine learning algorithms often work well with small and balanced datasets only [11]. Furthermore, feature engineering or extensive feature selection is essential for designing or selecting the most important feature set for model training. To lessen these difficulties, deep learning is an alternative solution. Deep learning has been used successfully in many areas of research, especially in health informatics [12–15], bioinformatics [16–19], and modern drug discovery [20–22].

In our study, we propose an effective computational model called iR1mA-LSTM to predict 1mA sites in human transcriptomes using attention-based long short-term memory (LSTM) networks and a one-hot encoding scheme. The RNA sequences collected from Chen et al.'s study [9] were used to develop and evaluate the model. To fairly assess the performance of iR1mA-LSTM, we compared it with RAMPred [8], iRNA-3typeA [9], and ISGm1A [10], which are the most relevant methods to our study.

2 Materials and Methods

2.1 Benchmark Dataset

The benchmark dataset for model development and evaluation was collected from Chen et al.'s study [9]. According to the information about the dataset, 1mA RNA sequences were obtained by mapping the peak signals of 1mA detected by an MeRIP-seq analyzer to the human genomes [23]. The adenines at 1mA sites are denoted as **A***, while adenines at non-1mA sites are denoted as **A**. To reduce sequence redundancy and similarity, sequences having more than 80% similarity were removed using the CD-HIT software [24]. After filtering these sequences, 6366 valid sequences were obtained and assigned as 'positive samples'. Each positive sample has a length of 41bp with **A*** located at the central position. To acquire the negative samples, RNA regions containing **A** but not showing any peak signal of 1mA on the MeRIP-seq analyzer were collected and truncated to form 41bp-length **A**-central sequences. Only 6366 negative sequences were randomly selected from a great number of the created **A**-central sequences. Finally, the benchmark dataset was constructed as a

Table 1 Data for model development and evaluation

Dataset	Number of samples		
	Positive	Negative	Total
Training	4870	4869	9739
Validation	541	542	1083
Independent test	955	955	1910

balanced dataset with 6366 samples for each class. In Chen et al.'s work, they used this benchmark dataset for model training and validation but not for model testing. In our work, we decided to split this benchmark dataset into three datasets: a training set, a validation set, and an independent test set, accounting for about 75%, 10%, and 15% of the dataset, respectively (Table 1).

2.2 *Sequence Embedding*

Figure 1 describes the sequence embedding process. Initially, a window of 3 was used to capture every three consecutive nucleic acids on 41bp-length sequences with a step of 1. For each sequence, a vector of 39 triplet keys was obtained. The combination of four different nucleic acids creates $4 \times 4 \times 4 = 64$ triplet keys, and these keys are indexed from 1 to 64. Index 0 was used for padding only. Sequence vectors of 39 triplet keys were then converted into corresponding sequence index vectors of size 41, including 2 auto-padding positions. The index vectors are the model inputs to enter an embedding layer. The embedding layer is a '*lookup table*' to return an embedding vector of size k for a specific triplet key. The embedding vectors of all triplet keys were randomly initialized and iteratively updated via training.

2.3 *Model Architecture*

Figure 2 illustrates the model architecture of our proposed method. The model architecture includes one Embedding layer, two bidirectional long short-term memory (Bi-LSTM) layers, an attention network, and two fully connected (FC) layers. The index vectors sized 1×41 are the model inputs that first enter the embedding layer of size $k = 100$ and are transformed to matrices sized 41×100. The embedding matrices sized 41×100 are then passed through two Bi-LSTM layers before entering the attention network. The attention network returns 1×100 matrices, which are then flattened before going through two FC layers. The FC1 and FC2 layers are activated by the ReLU and Sigmoid functions. The prediction threshold was set at 0.5. The loss function used was the binary cross-entropy:

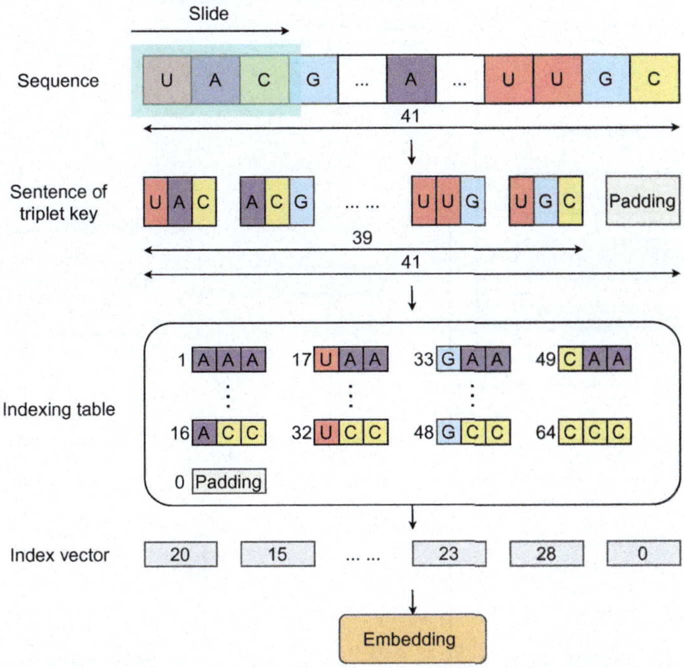

Fig. 1 Sequence embedding

$$Loss = \sum_{i=1}^{n} y_i \log \hat{y}_i + (1 - y_i) \log(1 - \hat{y}_i), \tag{1}$$

where y is the true label, and $\hat{y}$ is the predicted probability. The validation set was used to determine the stopping epoch for the model. The stopping epoch was determined at which the validation loss was minimum. The Adam optimization algorithm [25] and a learning rate of 0.0001 were used to optimize the model with every minibatch of 64 samples.

2.4 *Attention Network*

The attention network is included in Fig. 2. After leaving the second Bi-LSTM layer, hidden vectors sized 4×100 and output vectors sized 41×200 enter the attention network. Hidden vectors are split and summed on columns to form matrices sized 1×100 to enter the FC layer. The output vectors are split into forward and reverse output matrices for summation. The summed matrices M_{summed} sized 41×100 are activated by the Tanh function and transposed into matrices sized 100×41. The FC layer's outputs sized 1×100 are multiplied by the transposed matrices sized 100

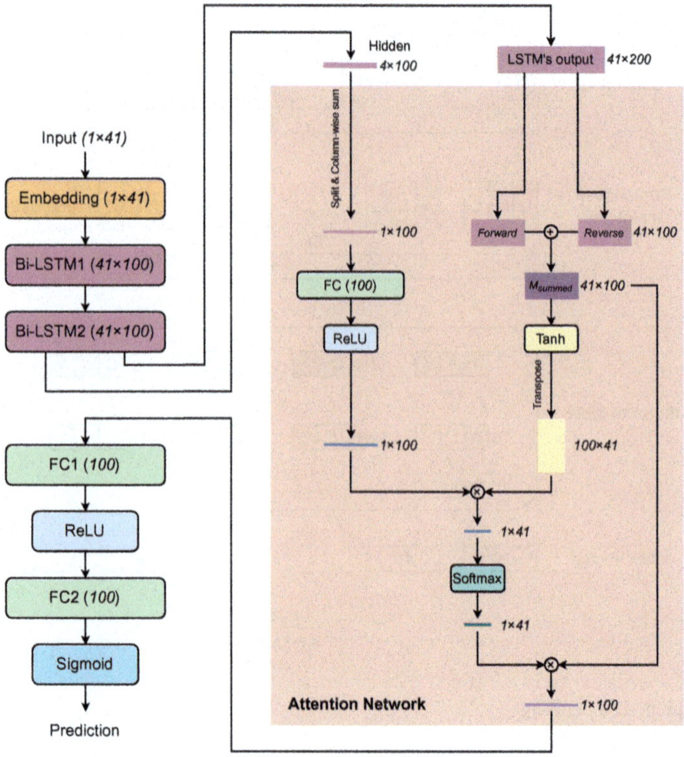

Fig. 2 Model architecture

× 41 to create matrices sized 1 × 41 which are activated by the Softmax function afterward. These matrices are finally multiplied by the M_{summed} matrices to create attention matrices sized 1 × 100. The attention matrices are then flattened before going to the next FC layer.

2.5 *Evaluation Metrics*

Multiple metrics, including the balanced accuracy (BA), sensitivity (SN), specificity (SP), precision (PR), F1-score, Matthews's correlation coefficient (MCC), the area under the receiver operating characteristic curve (AUCROC), and the area under the precision-recall curve (AUCPR), were measured to evaluate the model performance. TP, FP, TN, and FN are abbreviated terms for True Positives, False Positives, True Negatives, and False Negatives, respectively. The mathematical formulas for these metrics are expressed below.

$$BA = \frac{SN + SP}{2}, \quad (2)$$

$$SN = \frac{TP}{TP + FN}, \quad (3)$$

$$SP = \frac{TN}{TN + FP}, \quad (4)$$

$$PR = \frac{TP}{TP + FP}, \quad (5)$$

$$F1 = 2 \times \frac{PR \times SN}{PR + SN}, \quad (6)$$

$$MCC = \frac{TP \times TN - FP \times FN}{\sqrt{(TP + FP)(TP + FN)(TN + FP)(TN + FN)}}. \quad (7)$$

3 Results and Discussion

3.1 Model Evaluation

In our experiments, iR1mA-LSTM was implemented using PyTorch 1.12.0 and developed on Google Colab equipped with 25 GB of RAM and one NVIDIA Tesla T4 GPU. iR1mA-CNN was trained over 50 epochs. It took about 0.8 s to finish training one epoch and 0.25 s to finish model testing. Results indicate that our method obtained high performance with both AUCROC and AUCPR values of over 0.99 on the independent test set. Other metrics, including balanced accuracy, sensitivity, specificity, precision, MCC, and F1 score, are also greater than 0.95. The method was also examined in terms of robustness and stability by repeating our experiment in 10 trials. In each trial, the training, validation, and test sets were resampled to assure the randomness of data distribution. The ratio of the positive over negative samples in the training, validation, and test sets may be slightly different depending on data resampling. The results of the repeated experiments confirm the robustness and stability of iR1mA-LSTM in determining 1mA sites in RNA sequences (Table 2).

3.2 Comparative Analysis

Table 3 compares the model performance of iR1mA-LSTM with that of RAM-Pred [8], iRNA-3typeA [9], and ISGm1A [10]. The other three models were reimplemented based on their model descriptions. The decision threshold was set at 0.5

Table 2 Model performance of iR1mA-LSTM on the independent test set over ten trials

Trial	AUCROC	AUCPR	BA	SN	SP	PR	MCC	F1
1	0.9940	0.9955	0.9764	0.9675	0.9853	0.9851	0.9530	0.9762
2	0.9952	0.9966	0.9827	0.9738	0.9916	0.9915	0.9656	0.9826
3	0.9924	0.9950	0.9817	0.9728	0.9906	0.9904	0.9635	0.9815
4	0.9915	0.9944	0.9796	0.9686	0.9906	0.9904	0.9594	0.9794
5	0.9959	0.9971	0.9796	0.9780	0.9812	0.9811	0.9592	0.9795
6	0.9929	0.9951	0.9812	0.9717	0.9906	0.9904	0.9625	0.9810
7	0.9949	0.9961	0.9775	0.9780	0.9770	0.9770	0.9550	0.9775
8	0.9965	0.9976	0.9880	0.9812	0.9948	0.9947	0.9760	0.9879
9	0.9942	0.9958	0.9796	0.9728	0.9864	0.9862	0.9593	0.9794
10	0.9962	0.9971	0.9827	0.9801	0.9853	0.9853	0.9655	0.9827
Mean	0.9944	0.9960	0.9809	0.9745	0.9873	0.9872	0.9619	0.9808
SD	0.0017	0.0011	0.0032	0.0047	0.0054	0.0053	0.0065	0.0033

Table 3 Model performance of iR1mA-LSTM compared to other methods on the independent test set

Model	AUCROC	AUCPR	BA	SN	SP	PR	MCC	F1
RAMPred	0.9726	0.9768	0.9131	0.8984	0.9277	0.9256	0.8265	0.9118
iRNA-3typeA	0.9709	0.9757	0.9115	0.8963	0.9267	0.9244	0.8234	0.9102
ISGm1A	0.9836	0.986	0.9377	0.8754	**1.000**	**1.000**	0.8823	0.9336
iR1mA-LSTM	**0.9940**	**0.9955**	**0.9764**	**0.9530**	0.9529	0.9675	**0.9853**	**0.9851**

for all the models. The training, validation, and independent test sets in trial 1 were selected to develop all the models. Since the RAMPred, iRNA-3typeA, and ISGm1A models were developed using classical machine learning algorithms, the training and validation sets were merged to form only one training set. This new training set was then used in five-fold cross-validation to search for the best hyperparameters for these three models. The results show that our method outperformed the other three methods. iR1mA-LSTM ranked first, followed by ISGm1A, RAMPred, and iRNA-3typeA. The performance of iR1mA-LSTM exceeds that of the other methods, with both AUCROC and AUCPR values of over 0.99 (Fig. 3).

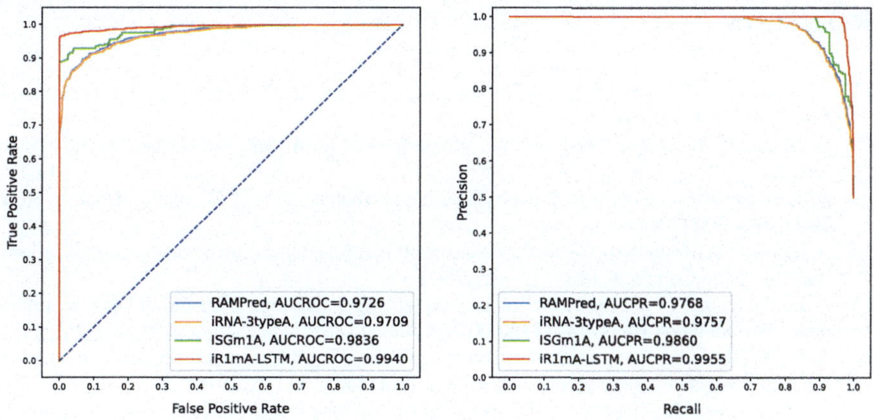

(a) ROC curves (b) PR curves

Fig. 3 ROC and PR curves of iR1mA-LSTM and other existing methods

3.3 Software Availability

To support researchers in detecting 1mA sites, we deployed i1mA-LSTM as a web server, which is available at https://github.com/mldlproject/2022-iR1mA-LSTM. iR1mA-LSTM assists users in detecting 1mA sites in human transcriptomes without dealing with mathematical difficulties. Users can do their prediction tasks with iR1mA-LSTM by following the simple steps that are written on the web server.

4 Conclusions

In this study, we introduce an advanced *in silico* approach to predict 1mA sites in human transcriptomes using bidirectional long short-term memory networks in combination with an attention network. Experimental results show that iR1mA-LSTM is an efficient, robust, and stable computational model. Our comparative analysis also confirms that iR1mA-LSTM worked better than other existing methods on the same independent test set. Our online web server was made with an easy-to-use interface and clear instructions to help users run prediction tasks quickly and easily.

Acknowledgements This project was supported in part by the Whitireia and WelTec Contestable fund.

References

1. R.J. Ontiveros, J. Stoute, K.F. Liu, The chemical diversity of RNA modifications. Biochem. J. **476**(8), 1227–45 (2019)
2. D. Wiener, S. Schwartz, The epitranscriptome beyond m6A. Nat. Rev. Genet. **22**(2), 119–31 (2021)
3. D.B. Dunn, The occurence of 1-methyladenine in ribonucleic acid. Biochem. Biophys. Acta. **46**(1), 198–200 (1961)
4. B.S. Zhao, I.A. Roundtree, C. He, Post-transcriptional gene regulation by mRNA modifications. Nat. Rev. Mol. Cell Biol. **18**(1), 31–42 (2017)
5. X. Li, X. Xiong, C. Yi, Epitranscriptome sequencing technologies: decoding DNA modifications. Nat. Methods **14**(1), 23–31 (2017)
6. X. Li, X. Xiong, K. Wang et al., Transcriptome-wide mapping reveals reversible and dynamic N1-methyladenosine methylome. Nat. Chem. Biol. **12**(5), 311–6 (2016)
7. X. Li, X. Xiong, M. Zhang et al., Base-resolution mapping reveals distinct m1A methylome in nuclear-and mitochondrial-encoded transcripts. Mol. Cell **68**(5), 993–1005 (2017)
8. W. Chen, P. Feng, H. Tang et al., RAMPred: identifying the N1-methyladenosine sites in eukaryotic transcriptomes. Sci. Rep. **6**(1), 1–8 (2016)
9. W. Chen, P. Feng, H. Yang et al., iRNA-3typeA: identifying three types of modification at RNA's adenosine sites. Molec. Therapy-Nucleic Acids **11**, 468–74 (2018)
10. L. Liu, X. Lei, J. Meng et al., ISGm1A: Integration of sequence features and genomic features to improve the prediction of human m1A RNA Methylation sites. IEEE Access. **8**, 81971–7 (2020)
11. T.H. Nguyen-Vo, Q.H. Nguyen, T.T.T. Do et al., iPseU-NCP: Identifying RNA pseudouridine sites using random forest and NCP-encoded features. BMC Genom. **20**(10), 1–11 (2019)
12. B.P. Nguyen, H.N. Pham, H. Tran et al., Predicting the onset of type 2 diabetes using wide and deep learning with electronic health records. Comput. Methods Programs Biomed. **182**, 105055 (2019)
13. Q.H. Nguyen, B.P. Nguyen, S.D. Dao et al., Deep learning models for tuberculosis detection from chest X-ray images, in *The 26th International Conference on Telecommunications (ICT 2019)* (IEEE, Hanoi, Vietnam, 2019), pp. 381–385
14. Q.H. Nguyen, R. Muthuraman, L. Singh et al., Diabetic retinopathy detection using deep learning, in *Proceedings of the 4th International Conference on Machine Learning and Soft Computing (ICMLSC 2020). ICPS* (ACM, Haiphong, Vietnam, 2020), pp. 103–107
15. Q.H. Nguyen, B.P. Nguyen, T.B. Nguyen et al., Stacking segment-based CNN with SVM for recognition of atrial fibrillation from single-lead ECG recordings. Biomed. Signal Process. Control **68**, 102672 (2021)
16. B.P. Nguyen, Q.H. Nguyen, G.N. Doan-Ngoc et al., iProDNA-CapsNet: identifying protein-DNA binding residues using capsule neural networks. BMC Bioinform. **20**(23), 1–12 (2019)
17. Q.H. Nguyen, T.H. Nguyen-Vo, N.Q.K. Le et al., iEnhancer-ECNN: identifying enhancers and their strength using ensembles of convolutional neural networks. BMC Genom. **20**(9), 1–10 (2019)
18. N.Q.K. Le, B.P. Nguyen, Prediction of FMN binding sites in electron transport chains based on 2-D CNN and PSSM Profiles. IEEE/ACM Trans. Comput. Biol. Bioinf. **18**(6), 2189–97 (2019)
19. N.Q.K. Le, Q.H. Nguyen, X. Chen et al., Classification of adaptor proteins using recurrent neural networks and PSSM profiles. BMC Genom. **20**(Suppl 9), 1–9 (2019)
20. T.H. Nguyen-Vo, L. Nguyen, N. Do et al., Predicting drug-induced liver injury using convolutional neural network and molecular fingerprint-embedded features. ACS Omega **5**(39), 25432–9 (2020)
21. T.H. Nguyen-Vo, Q.H. Trinh, L. Nguyen et al., Predicting antimalarial activity in natural products using pretrained bidirectional encoder representations from transformers. J. Chem. Inform. Model. (2021)

22. T.H. Nguyen-Vo, Q.H. Trinh, L. Nguyen et al., iCYP-MFE: identifying human cytochrome P450 inhibitors using multitask learning and molecular fingerprint-embedded encoding. J. Chem. Inform. Model. (2021)
23. D. Dominissini, S. Nachtergaele, S. Moshitch-Moshkovitz et al., The dynamic N1-methyladenosine methylome in eukaryotic messenger RNA. Nature **530**(7591), 441–6 (2016)
24. Y. Huang, B. Niu, Y. Gao et al., CD-HIT Suite: a web server for clustering and comparing biological sequences. Bioinformatics **26**(5), 680–2 (2010)
25. D.P. Kingma, J. Ba, Adam: a method for stochastic optimization. arXiv (2014)

Explaining an Empirical Formula for Bioreaction to Similar Stimuli (Covid-19 and Beyond)

Olga Kosheleva, Vladik Kreinovich, and Nguyen Hoang Phuong

Abstract A recent comparative analysis of biological reaction to unchanging versus rapidly changing stimuli—such as Covid-19 or flu viruses—uses an empirical formula describing how the reaction to a similar stimulus depends on the distance between the new and original stimuli. In this paper, we provide a from-first-principles explanation for this empirical formula.

1 Formulation of the Problem

Bioreactions: general reminder. Most living creatures have the ability to learn. When we first encounter some stimulus—e.g., some chemical substance or some bacteria—we do not know whether this stimulus is harmful or beneficial. This encounter—and several similar encounters—shows us whether this particular stimulus is harmful, beneficial, or neutral.

We learn from this experience, so next time, when we encounter a similar stimulus, we know how to react: e.g., fight or flee if this stimulus is harmful.

Bioreaction depends on whether stimuli evolve with time. Some stimuli—e.g., smells associated with some chemicals—do not change with time. So, we learn to associate the smell with the corresponding stimulus—e.g., the smell of a dangerous

O. Kosheleva
Department of Teacher Education, University of Texas at El Paso, 500 W. University, El Paso, TX 79968, USA
e-mail: olgak@utep.edu

V. Kreinovich (✉)
Department of Computer Science, University of Texas at El Paso, 500 W. University, El Paso, TX 79968, USA
e-mail: vladik@utep.edu

N. H. Phuong
Artificial Intelligence Division, Information Technology Faculty, Thang Long University, Nghiem Xuan Yem Road, Hoang Mai District, Hanoi, Vietnam

N. H. Phuong and V. Kreinovich (eds.), *Deep Learning and Other Soft Computing Techniques*, Studies in Computational Intelligence 1097,
https://doi.org/10.1007/978-3-031-29447-1_6

predator with danger, and the smell of juicy edible apples or mushrooms with tasty food. Living creatures can become very selective in this association, easily distinguishing the smell of a dangerous wolf from a similar smell of a friendly dog. In such situation, an optimal strategy for a living creature would be to remember the exact stimulus—and only react to exactly this stimulus.

Other stimuli vary in our lifetime. For example, many viruses—e.g., flu and Covid-19 viruses—evolve every year. In this case, if the cells protecting our bodies from these viruses would only react to the exact shape of the viruses they encountered last year, this would leave us unprotected against even a very minor virus mutation. In such cases, it is important to react not only to the exact same stimulus as before, but also to stimuli which are similar to the ones that we previously encountered.

The closer the new stimulus to the original one, the stronger the reaction. When we encounter the exact same dangerous stimulus, we are absolutely sure that this stimulus is dangerous, so we should react with full force. On the other hand, when we encounter a stimulus which is similar to the original stimulus, we are no longer 100% sure: this new stimulus may be harmless and we may be wasting resources if we immediately launch a full-blown attack against it—resources that can be needed in the future, when a serious danger comes.

So, the father away the new stimulus from the original one, the weaker should be the bioreaction to this stimulus—and, vice versa, the closer the new stimulus to the original dangerous one, the stronger should the bioreaction be.

An empirical formula describing this dependence. In many biological situations, there is a natural way to measure the distance d between two stimuli—e.g., we can measure the distance between the two DNAs by the total length of the parts which are specific to one of them.

Recent papers [3, 4] have shown that the observations are in good accordance with the following dependence of the reaction force f on the distance d:

$$f = F_0 \cdot \exp\left(-k \cdot d^{\theta}\right), \tag{1}$$

for some parameters $F_0 > 0$, $k > 0$, and $\theta > 0$—and these papers show that the observed biological values of these parameters are close to optimal.

Problem. A natural question is: how to explain this empirical dependence?

What we do in this paper. In this paper, we provide a possible from-first-principles explanation for the empirical dependence (1).

2 General Idea Behind Many from-First-Principles Explanations

Before we consider this specific problem, let us recall where many from-first-principles explanations come from.

Numerical values versus actual values. What we want is to find dependence between the *actual* values of the corresponding quantities. However, all we can do is come up with relation between *numerical* values describing these properties.

Numerical values depend not only on the quantity itself, they also depend on the choice of the measurement scale. For example, the numerical values depend on the choice of the measuring unit. If we replace the original measuring unit with the one which is λ times smaller, then all numerical values multiply by λ: $x \mapsto \lambda \cdot x$. In particular, if we use centimeters instead of meters, then 1.7 m becomes 170 cm.

For many physical quantities like time and temperature, the numerical values also depend on the selection of the starting point. If instead of the original starting point, we choose a news starting point which is x_0 units earlier, then all numerical values are changed: $x \mapsto x + x_0$.

There may also be non-linear rescalings. In all these cases, moving to a different scale changes the numerical value, from the original numerical value x to the new value $T_c(x)$, where c is the parameter, and $T_c(x)$ is the corresponding transformation. For example, for changing the measuring unit, $T_c(x) = c \cdot x$, for changing the starting point, $T_c(x) = x + c$. etc.

Invariance: general idea. In many practical situations, there is no meaningful way to select a scale, all scales are equally reasonable. In such situations, it makes sense to require that the relation $y = f(x)$ between the two quantities x and y has the same form in all these scales.

Of course, if we re-scale x, i.e., replace it with $x' = T_c(x)$, then, to preserve the relation between x and y, we also need to re-scale y, i.e., to apply an appropriate transformation $y \mapsto y' = T'_{c'}(y)$. Then, we can require that for every c there exists a c' for which $y = f(x)$ implies that for $x' = T_c(x)$ and $y' = T'_{c'}(y)$, we have $y' = f(x')$.

Invariance: example. For example, the formula $a = s^2$ relating the square's area a with its side a remains valid if we replace meters with centimeters, but then, we need to correspondingly replace square meters with square centimeters. In this case, for $T_c(x) = c \cdot x$, we have $T'_{c'}(y) = c' \cdot y$ with $c' = c^2$.

How invariance explains a dependence: example. Let us consider situations when for every c, there exists a value $c'(c)$ (depending on c) for which $y = f(x)$ implies $y' = f(x')$, where $x' = c \cdot x$ and $y' = c'(c) \cdot y$.

Substituting the expressions for x' and y' into the formula $y' = f(x')$ and taking into account that $y = f(x)$, we conclude that for every x and c, we have $f(c \cdot x) = c'(c) \cdot f(x)$. It is known (see, e.g., [1]) that every continuous (even every measurable) solution to this functional equation has the form

$$y = A \cdot x^b$$

for some A and b.

Thus, this ubiquitous *power law* can be explained by the corresponding invariance.

3 Let Us Apply This General Idea to Our Problem

For our problem, what are the natural scales? To apply the above general ideas to our problem—of finding the dependence between the interaction force f and the distance d—we need to understand what are the natural scales for measuring these two quantities: distance d and force f.

Case of distance. For distance, the usual distance measures are appropriate. So, a natural change in scale in the change of the measuring unit $d \mapsto c \cdot d$.

Case of force: analysis of the problem. On the other hand, for force, the situation is not that straightforward.

In a purely mechanical environment, we can combine several forces together, so we can easily see what corresponds to 2 or 3 unit forces. So, if we select a unit force f_0, we can talk about the force $2 f_0$ which is equivalent to a joint action of two unit forces, the force $3 f_0$ which is equivalent to a joint action of three unit forces, etc. In such an environment, the following will be a natural scale for measuring force: the numerical value of the force f is the number n for which the force f is equivalent to the joint action of n unit forces $f \approx n \cdot f_0$, i.e., in effect, the value $n \approx f/f_0$.

However, for biosystems, no such natural combination of forces is possible. The only thing we can do is compare two forces. Of course, if the forces are almost the same, we will not be able to distinguish them. So, if we select a unit force f_0, then the next natural value f_1 is the smallest value $f_1 > f_0$ that can be distinguished from f_0. After that, the next natural value f_2 is the smallest value $f_2 > f_1$ that can be distinguished from f_1, etc.

Let us describe the above idea in precise terms. To describe these values in precise terms, we need to be able to determine, for each force f, the smallest value $g = F(f) > f$ which can be distinguished from f.

Processes involving forces do not depend on the exact choice of the physical measuring unit for a force. So, if we have $g = F(f)$ in the original units for physical force, then in a new scale, for $f' = c \cdot f$ and $g' = c \cdot g$, we should have $g' = F(f')$.

Substituting the above expressions for f' and g' into this formula and taking into account that $g = F(f)$, we conclude that $F(c \cdot f) = c \cdot F(f)$. In particular, for $f = 1$, we get $F(c) = q \cdot c$, where we denoted $q \stackrel{\text{def}}{=} F(1)$. Thus, we have $f_1 = q \cdot f_0$, $f_2 = q \cdot f_1 = q^2 \cdot f_0$, $f_3 = q \cdot f_2 = q^2 \cdot f_0$, and, in general, $f_n = q^n \cdot f_0$.

So, a natural scale for measuring the bioforce f is the number n for which f corresponds to the n-th element on this scale, i.e., for which $f \approx q^n \cdot f_0$ and

$$n \approx \log_q(f/f_0) = \frac{\ln(f/f_0)}{\ln(q)}. \quad (2)$$

Comment. It should be mentioned that the formula (2) describes what is known in physiology as Weber-Fechner Law—that the intensity of each sensation is proportional to the logarithm of its physical measure (energy or force); see, e.g., [2].

From the above somewhat simplified description to a more realistic one. In the above analysis, we implicitly assumed that for every two forces, we can either distinguish them or not. However, this implicit assumption is a simplification.

When one of the forces is much larger than the other one, then, of course, this is absolutely true. However, as the forces get closer to each other, there appears a probability that we will not be able to distinguish these forces—and the closer these forces to each other, the larger this probability. When the compared forces are very close, this probability becomes so large that, for all practical purposes, we cannot distinguish them.

In view of this fact, to describe the scale, we need to also select a confidence level with which we can distinguish the two forces. If we select this confidence level too high, then we will need a large value q—the ratio of the forces f_1/f_0. The smaller q, the smaller the confidence level.

We arrive at different natural measurement scales for (bio)force. There is no fixed confidence level, so there is no preferred value q. In other words, we can have different natural scales of type (2) corresponding to different values q.

What is the transformation between two different natural scales for measuring (bio)force? Suppose that in addition to the scale (2) that corresponds to some value q, we also have a different scale

$$n' \approx \frac{\ln(f/f_0)}{\ln(q')} \quad (3)$$

corresponding to a different value q'. How can we transform the value n corresponding to a given physical force f on a q-based scale into a value n' corresponding to the same force f on a different q'-based natural scale?

By comparing the formulas (2) and (3), one can see that the relation between n' and n has a very simple form: $n' = c \cdot n$, where we denoted

$$c \stackrel{\text{def}}{=} \frac{\ln(q)}{\ln(q')}.$$

Which dependencies are invariant with respect to these transformations. We want to find out how force f depends on the distance d. For biosystems, a natural way to describe this dependence is by using natural scale n for bioforce. Thus, we want to describe how the bioforce n depends on the distance.

Both for the distance and for the bioforce, natural transformations have the form $x \mapsto c \cdot x$. Thus, a natural invariance of the dependence $n = N(d)$ means that for every c, there should exist some value $c'(c)$ such that $n = N(d)$ implies that $n' = N(d')$, where we denoted $d' \stackrel{\text{def}}{=} c \cdot d$ and $n' \stackrel{\text{def}}{=} c'(c) \cdot n$. We have already mentioned that this invariance implies that

$$n = A \cdot d^b \tag{4}$$

for some constants A and b.

This explains the desired dependence between f and d. Our ultimate objective is to explain the empirical dependence (1) between the physical force f and the distance d. Let us therefore see how the dependence (4) between n and d will look like in terms of the dependence between f and d.

To find this out, let us plug in the expression (4) for n into the above formula $f = f_0 \cdot q^n$, i.e., equivalently, $f = f_0 \cdot \exp(n \cdot \ln(q))$. This substitution leads to

$$f = f_0 \cdot \exp(A \cdot \ln(q) \cdot d^b).$$

This is exactly the formula (1), for $F_0 = f_0$, $k = -A \cdot \ln(q)$, and $\theta = b$.

Thus, we indeed have a from-first-principles explanation for the above empirical dependence.

Acknowledgements This work was supported in part by the National Science Foundation grants 1623190 (A Model of Change for Preparing a New Generation for Professional Practice in Computer Science), and HRD-1834620 and HRD-2034030 (CAHSI Includes), and by the AT&T Fellowship in Information Technology.

It was also supported by the program of the development of the Scientific-Educational Mathematical Center of Volga Federal District No. 075-02-2020-1478, and by a grant from the Hungarian National Research, Development and Innovation Office (NRDI).

References

1. J. Aczél, J. Dhombres, *Functional Equations in Several Variables* (Cambridge University Press, 2008)
2. J.T. Cacioppo, L.G. Tassinary, G.G. Berntson (eds.), Handbook of Psychophysiology. (Cambridge University Press, Cambridge, 2019)
3. O.H. Schnaack, A. Nourmohammad, Optimal evolutionary decision-making to store immune memory. eLife **10**, Paper e61346 (2021)
4. O.H. Schnaack, L, Peliti, A. Nourmohammad, *Learning and organization of memory for evolving patterns* (2021), arXiv:2106.02186v1

Game-Theoretic Approach Explains—On the Qualitative Level—The Antigenic Map of Covid-19 Variants

Olga Kosheleva, Vladik Kreinovich, and Nguyen Hoang Phuong

Abstract To effectively defend the population against future variants of Covid-19, it is important to be able to predict how it will evolve. For this purpose, it is necessary to understand the logic behind its evolution so far. At first glance, this evolution looks random and thus, difficult to predict. However, we show that already a simple game-theoretic model can actually explain—on the qualitative level—how this virus mutated so far.

1 Formulation of the Problem

An important problem. Like many viruses, the virus that causes Covid-19 rapidly evolves, so that vaccines which are very efficient for the original variants are not as efficient for the new variants. To be better prepared for the future variants, it is desirable to predict how the virus will evolve in the future. To be able to do it, we need to understand how (and why) it evolved the way it did.

The problem looks complicated. A recent map [1, 2] provides a visual 2-D description of the evolution of the main Covid-19 variants, from the original Alpha to Beta, Gamma, Delta, and Omicron. At first glance, the changes look somewhat random and chaotic, leaving us with an impression that probably no reasonable predictions are possible.

O. Kosheleva
Department of Teacher Education, University of Texas at El Paso, 500 W. University, El Paso, TX 79968, USA
e-mail: olgak@utep.edu

V. Kreinovich (✉)
Department of Computer Science, University of Texas at El Paso, 500 W. University, El Paso, TX 79968, USA
e-mail: vladik@utep.edu

N. H. Phuong
Artificial Intelligence Division, Information Technology Faculty, Thang Long University, Nghiem Xuan Yem Road, Hoang Mai District, Hanoi, Vietnam

N. H. Phuong and V. Kreinovich (eds.), *Deep Learning and Other Soft Computing Techniques*, Studies in Computational Intelligence 1097,
https://doi.org/10.1007/978-3-031-29447-1_7

What we do in this paper. In this paper, we show that, at least on the qualitative level, natural game-theoretic ideas explain the current evolution—and thus, hopefully enable us to predict the direction of future changes.

2 A Simplified Game-Theoretic Model and the Resulting Explanation

Main idea. As the virus starts affecting the population, people's bodies start developing antibodies to the virus's original version. So, to stay effective, the virus has to mutate. The mutated version also causes the bodies to develop protection, so further mutations are needed.

The larger the distance between the two variants A and B, the less effective A-caused antibodies against the variant B. Thus, from the virus's viewpoint, after variants A_1, ..., A_n, it makes sense to select, as the next variant, the variant A for which the smallest of the distances $\min(d(A, A_1), \ldots, d(A, A_n))$ is the largest possible—this will guarantee that the new variant will be the most effective against all A_i-produced antibodies.

If there are several variants A with this largest value, then, for the virus, it is reasonable to select the variant for which the second smallest of the distances $d(A, A_i)$ is the largest, etc.

Comment. Of course, the virus is not an intelligent being, it does not directly select its next mutation: mutations happen randomly, some lead to more effective variants, some to less effective ones, and the most effective one becomes dominant. In this sense, after the variants A_1, ..., A_n, the next effective one—the next dominant variant—is the one for which the value $\min(d(A, A_1), \ldots, d(A, A_n))$ describing the variant's effectiveness is the largest.

Let us trace this idea on a simple geometric example. In the 2-D description, all possible variants belong to some reasonable planar area. Let us consider the simplest case, when this area is a disk:

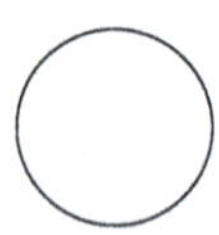

The first variant A_1 appears somewhere inside this disk-shaped area. For simplicity, let us assume that it is located in the center of the disk:

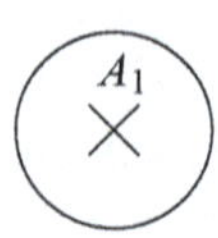

Following the above description, as the next variant A_2, we select the point in the disk for which the distance to the center is the largest possible. One can easily see that the largest possible distance is equal to the radius r of the disk, and this distance is attained at any point on the corresponding circle. So, the next variant is located on the circle that borders the disk:

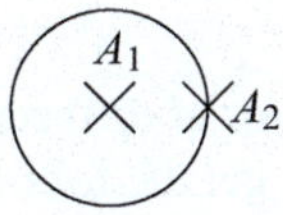

What happens next? The distance from any point A inside the disk to its center A_1 cannot exceed r. So the smallest distance from A to points A_1 and A_2 cannot be larger that r. So, ideally, as the next point A_3, we should select a point A for which this smallest distance $\min(d(A, A_1), d(A, A_2))$ is equal to the largest possible value r. This mean, in particular, that the distance $d(A, A_1)$ is at least r—and since this distance cannot exceed r, this means that it must be exactly equal to r, i.e., that the point A_3 should also be located on the circle.

All the points A on the circle has the exact same distance $d(A, A_1)$. So, in line with the above description, of all points A from the circle, we should select the point for which the distance $d(A, A_2)$ is the largest possible. One can see that this largest distance—equal to $2r$—is attained when the point A_3 is on the same line as A_1 and A_2 but on the opposite side of A_1:

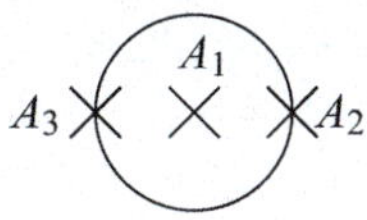

What now? Similarly to the previous case, we can conclude that the next point A_4 should also be located on the circle, and it should be selected in such a way that the smallest of the two distances $\min(d(A, A_2), d(A, A_3))$ should be the largest possible. One can check that this point should be located on the circle exactly in the middle between A_2 and A_3:

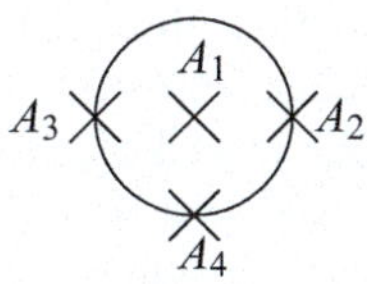

Similarly, the next point A_5 should be located on the same circle, also exactly in the middle between A_2 and A_3:

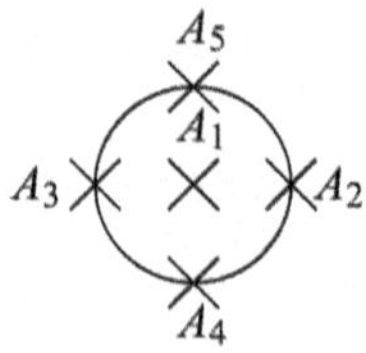

Let us map the consequent locations of different variants. If we map the locations of variants A_1 through A_4, then we get the following picture:

$$\begin{array}{ccccc} A_3 & - & A_1 & - & A_2 \\ & & | & & \\ & & A_4 & & \end{array}$$

If we add A_5, then we get the following:

$$\begin{array}{ccccc} & & A_5 & & \\ & & | & & \\ A_3 & - & A_1 & - & A_2 \\ & & | & & \\ & & A_4 & & \end{array}$$

On the qualitative level, the A_1-A_4 map is almost exactly the Covid-19 antigen map. Indeed, we start with the first variant Alpha (A_1 in our notations). Then, the variant Beta (A_2) is on one side of A_1, while the variant Delta (A_3) is approximately on the same line on the other side of A_1. (We skipped Gamma, which is exactly in between Beta and Delta on the antigen map, so it must simply be a transitional state.)

Then comes Omicron (A_4) which is obtained by moving in a different direction than before.

So what next? Our model's prediction is the next variant A_5 will be along the line A_1A_4, but on the other side of A_1 than Omicron (A_4).

Of course, this is a very crude model. The above model is very approximate. For example, in this model the distance from all three consequent variants to the original variant A_1 is the same, while in reality, the distance from Alpha (A_1) to Omicron (A_4) is much larger than the distances from Alpha to Beta and Delta.

However, the fact that this simple model explains the seemingly random changes gives us hope that a further development of this model can lead to quantitative explanations—and thus, more reliable predictions.

Acknowledgements This work was supported in part by the National Science Foundation grants 1623190 (A Model of Change for Preparing a New Generation for Professional Practice in Computer Science), and HRD-1834620 and HRD-2034030 (CAHSI Includes), and by the AT&T Fellowship in Information Technology.
It was also supported by the program of the development of the Scientific-Educational Mathematical Center of Volga Federal District No. 075-02-2020-1478, and by a grant from the Hungarian National Research, Development and Innovation Office (NRDI).

References

1. E. Waltz, The algorithm that mapped Omicron. IEEE Spectrum 26–31 (2022)
2. S.H. Wilks, B. Mühlemann, X. Shen, S. Türeli, E. B. LeGresley, A. Netzl, M.A. Caniza, J.N. Chacaltana-Huarcaya, X. Daniell, M.B. Datto, T.N. Denny, C. Drosten, R.A.M. Fouchier, P.J. Garcia, P.J. Halfmann, A. Jassem, T.C. Jones, Y. Kawaoka, F. Krammer, C. McDana, R. Pajon, V. Simon, M. Stockwell, H. Tang, H. van Bakel, R. Webby, D.C. Montefiori, D.J. Smith, Mapping SARS-CoV-2 antigenic relationships and serological responses (2022), bioRxiv preprint https://doi.org/10.1101/2022.01.28.477987, posted January 28, 2022

Application of the Artificial Intelligence Technique to Recognize and Analyze from the Image Data

Lu Anh Duy Phan and Ha Quang Thinh Ngo

Abstract Recently, there have been numerous applications of deep learning-based approaches such as speech recognition, face detection or risk prediction. This investigation introduces a novel approach to deploying the deep learning technique to evaluate physical gestures. First, the system's overall scheme and the local setup are discussed. Then, the scoring mechanism to reward is conceptually established to estimate human performance. This score is immediately demonstrated how hard to imitate the reference model. Later, some statistics indicators, for instance, timing, type of actions or emotion, could be visually analyzed and performed. The contributions of this paper are (1) to initially innovate the overall framework of the system, (2) to create the scoring mechanism to evaluate and (3) to denote the indicators to assess human performance. From these results, it has been seen that the proposed solution could be applied in teaching activities, sports analytics or gesture-based control.

Keywords Data analytics · Image processing · Motion control · Index system

1 Introduction

The analysis of sports activities has been considerably investigated in recent years as a growing investment from various sources such as government, local ministry and the private sector. This field requires multidisciplinary methods, including mechatronics with an imaginative approach, electrical engineering and computer science, sports, or statistics [1–4]. Although the increasing interest in bodily activities has caused the development of different methods to examine the physical parameters, this work still has some challenges to overcome. Activities analytics have gained much attention,

L. A. D. Phan · H. Q. T. Ngo (✉)
Faculty of Mechanical Engineering, Ho Chi Minh City University of Technology (HCMUT), 268 Ly Thuong Kiet, District 10, Ho Chi Minh, Vietnam
e-mail: nhqthinh@hcmut.edu.vn

Vietnam National University-Ho Chi Minh City (VNU-HCM), Linh Trung Ward, Thu Duc City, Ho Chi Minh, Vietnam

N. H. Phuong and V. Kreinovich (eds.), *Deep Learning and Other Soft Computing Techniques*, Studies in Computational Intelligence 1097,
https://doi.org/10.1007/978-3-031-29447-1_8

especially in the world of famous team sports such as cricket [5], soccer [6, 7], basketball [8], and volleyball [9]. The need to analyze physical games has been continuously augmented when a large proportion of residents changes their lifestyle.

In general, there are two kinds of statistical analyses to improve player performance: on-field/online and off-field/offline. Online research requires sophisticated techniques and advanced computing tools to measure performance indicators. Also, it needs a lot of operators and executives who simultaneously work on the site [10, 11]. Reversely, the offline method could spend less time and cost since the whole process of analysis focus on video. However, the quality and view of the video play a vital role in carrying out the best results [12–14]. In both cases, intending to enhance and exploit the upper limits of players as well as the efficiency of proper investment, these researches are helpful not only in the technical analysis but also in multivariate functions, for instance, tactics or team formulation.

For the present, to distinguish any relationship or interaction, enormous human labour must be involved in different roles. Various approaches are mentioned in this field to analyze and evaluate depending on the technique used. In [15], an index system to reflect the characteristics of volleyball was realized. The fuzzy evaluation was adopted to produce the weight assignment and pass the consistency test. The outcomes were provided for teachers to improve the teaching quality of physical education in colleges. In the same field, but using the other method, researchers [16] deployed the data fusion to recognize each volleyball motion accurately. According to the synchronization between recorded video and acceleration data from the player's wrist, several kinds of essential characteristics were identified by the neural network. It was then utilized to train the new player with high accuracy.

Consequently, those works were often initiated after the game or match ended. The real-time technique should be considered when the statistics analysis occurs when the competitive game is scoring. In [17], a real-time object detection system mainly integrating YOLO (You Only Look Once) was established to detect the presence of both objects and players and recognize the number of specific movements. Even under such poor conditions, for example, if the players move out of the frame due to the camera shift, this system could uninterruptedly track these players. In the equivalent technique, the lightweight model of YOLO and convolutional neural network without additional preprocessing [18] were encouraged to be set up. This development is significant for those with specific disabilities can communicate with others by using gestures.

From these reports, it could be considered that the applications of deep learning techniques are essential to analyze human gestures as an automated solution. In [19], a mobile application to monitor the heart rate during physical exercise was innovated to understand the effects of training intensity due to health status and performance. Also, this algorithm has been highly used in facial recognition and severe problems. In addition, for the mission of automatic person detection in the thermal image, a method using a convolutional neural network-based model was investigated [20]. The experiments were in three outdoor conditions: clear weather, rain and fog. The system performance was excellent regarding average precision for all validated scenarios. In the same approach, the extended version to detect humans via thermal image was presented in [21]. Compared to RGB images, these results were superior, faster and more precise than the conventional methods. The reminders of this paper are as follows. Section 2 describes the framework of the overall system and explains the hardware set-up and data collection. Section 3 illustrates the method to evaluate each gesture owing to the scoring mechanism. The experimental results are depicted in Sect. 4 to estimate the system performance. Finally, several conclusions are carried out in Sect. 5.

2 System Design

This paper proposes a novel approach to analyzing physical gestures using the deep learning technique. Firstly, a data set which consists of many exercises for youth or children was created to train. In this study, these lessons are classified into three levels beginner, intermediate and professional. Secondly, the integrated software for recording video, analyzing activities and the scoring mechanism is programmed. The sports players could revise their actions in each gesture along with several performance indicators. Last but not least, the scoring mechanism would release the estimated values compared to the reference model.

2.1 Framework of Overall System

Figure 1 describes the entire system using the deep learning technique. The sampling data is collected in advance to be used as inputs. In the training process, the convolutional layer is to compute with more acceptable feature maps combined to predict better. These features are repeated for the minimum factors, fusing them from previously convolutional layers at a reasonable scale. Three sets of box predictions are generated in the feature maps. The whole process is often done offline to provide the exact information for recognition. In experiments, a digital camera captures and transmits any action to the host computer. At this time, the gesture recognition model separates each step. Then, it labels a frame of the image by using the information from the training data set. The outputs of this routine are fed to the scoring mechanism to evaluate the performance of players.

2.2 Hardware Setup

In fact, the digital camera plays an important role for this investigation. It must be able to capture the high-quality of images to avoid blurring in poor conditions. Additionally, the depth map must be obtained since the image processing requires a lot of steps and data. To match with these requirements, a depth camera Realsense

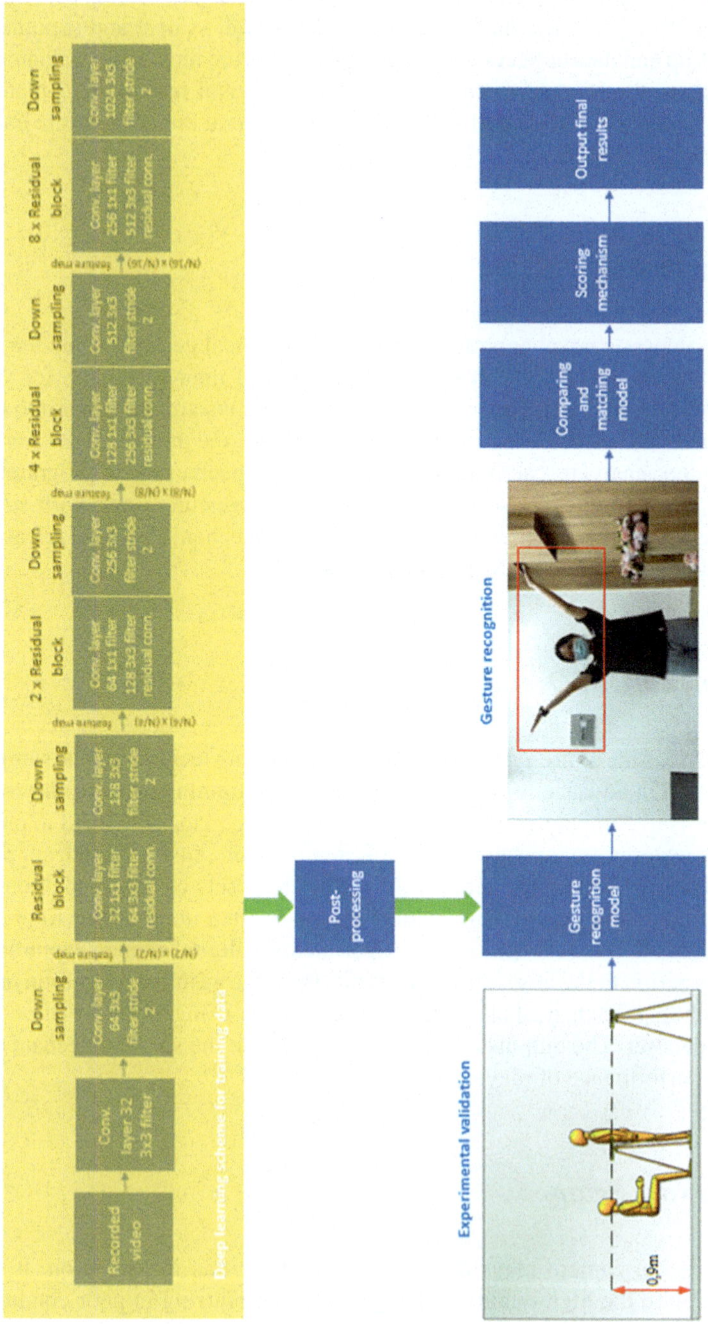

Fig. 1 Overall scheme of the proposed system

D435 in Fig. 2 is elected as main source. It has wide field of view, and is perfect for applications such as robotics or augmented and virtual reality, where seeing as much of the scene as possible is vitally important. With a range up to 10 m, this small form factor camera can be integrated into any solution with ease, and comes complete with our Intel RealSense SDK 2.0 and cross-platform support.

With the powerful camera, the setting height is direct and in the eye view of player. To avoid the loss of captured data in the event of being obscured, the second camera or preventive camera is placed at the side view. To illustrate the installation of the hardware system, Fig. 3 represents the estimated distance from human to camera while Fig. 4 demonstrates the overview of whole setup.

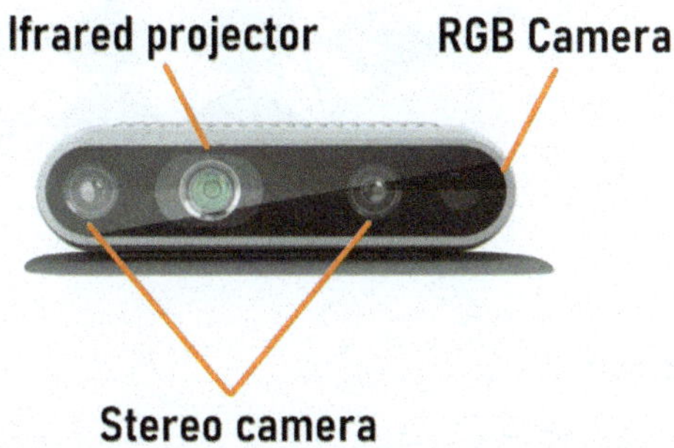

Fig. 2 Description of components in Realsense camera

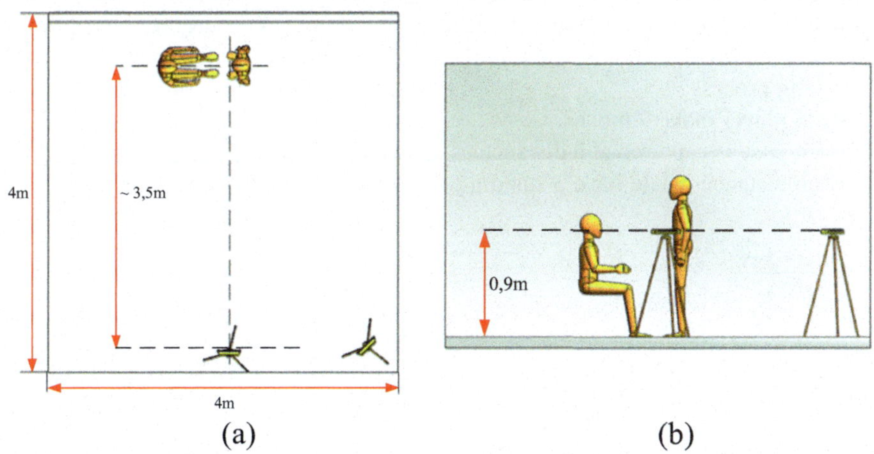

Fig. 3 Distancing estimation from camera to sport player, **a** top view, **b** side view

Fig. 4 Global view of on-site setup

2.3 Data Collection

In a level, there are many different gestures which may vary for each player. The guider or teacher who becomes a reference model would act firstly. Later, player follows these exercises and get a score. In order to better absorb, the movements would be very simple, including 9 positions and arranged according to each level. To collect data for the identification process, the recording work is according to the following process:

Each player make 9 moves.

Each pose is captured at a distance as shown in Fig. 5.

Each distance would have 3 standing positions: left, middle and right.

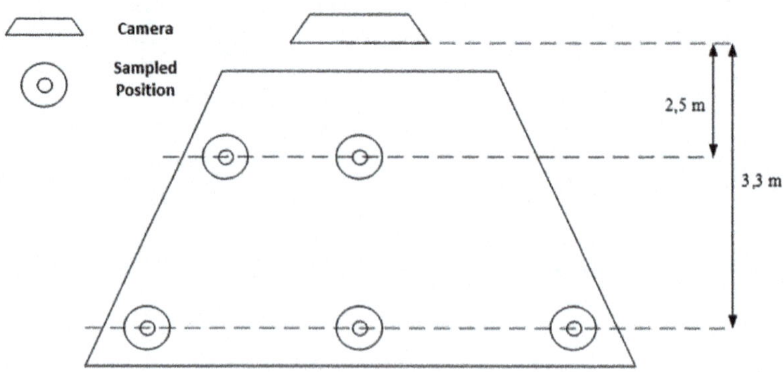

Fig. 5 Location of data collection

Each pose takes 3 angles: facing the camera, rotated left 30° and rotated right 30°. Each angle would capture 10 skeletons.

As a result, a dataset would include 180 frames for each pose and 8100 frames for the whole data set.

3 Scoring Mechanism

3.1 *Conceptual Definition*

Human posture can be recognized by information from color images taken by normal cameras. However, the main obstacle of traditional resolution methods is that extracting features from images captured by conventional cameras is still difficult due to noise, shooting angle, light, and environmental influences. Therefore, the depth camera could be used to help provide depth information and track the skeleton of the person standing in front of the camera.

From the above requirements, a common type of depth camera that used a lot in research articles on human posture recognition is presented with specifications as in Table 1.

The recorded data results in coordinates of 25 joints that determined in real time forming a skeleton map as shown in Fig. 6a. The coordinates of the joints are unique and can perfectly represent a movement.

3.2 *Vector of Limb*

In order for digital camera to capture and feedback the coordinates of the joints, so the limb vectors would be used to represent the skeleton posture of the person in front. A limb vector is defined as the connection between two joints in Fig. 6b. Like geometric vectors, the description of them denotes a single gesture without shifting

Table 1 The specification parameters of the proposed system

Parameters		Value
Depth sensor		Time of flight (ToF)
Resolution of RGB camera		1920 × 1080, 30 fps
Resolution of depth camera		512 × 424, 30 fps
Visual scope	RGB camera	70° × 60°
	Depth camera	84,1° × 53,8°
Operating range		0,5–4,5 m
Points of skeleton		25 points
Max number of human detection		6

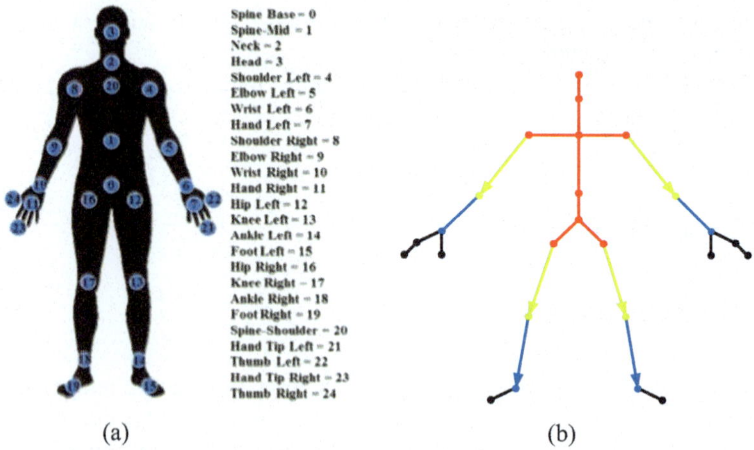

Fig. 6 Map of skeleton's positions (**a**), skeleton data and limbs vector (**b**)

or scaling. Body motion can also be perfectly simulated by the rotation of each vector around its original joint.

By considering the characteristics of human movement and body structure, it could be reduced the redundant data via analyzing the limb vectors in different groups. Vector of head-body which includes the head, shoulders and hips as Fig. 6b with red endpoints, has few individual and strong movements. The rotation and flexion of this body part is associated with the extremities. Therefore, these joints would not be comprised in the representative group.

Vector of level 1: includes elbow and thigh as Fig. 6b with yellow lines. This vector contains a lot of information of movements and gestures. Therefore, these vectors would be classified as representative groups.

Vector of level 2: includes arms and legs as Fig. 6b with blue lines. These vectors last longer than those of level 1, and make a significantly visual impression. Therefore, they could be classified as representative ones.

Hands and feet: camera can track feet and hands as Fig. 6b with black lines, but it is often unstable during acquisition. Additionally, the movements of hand and foot are often negligible in the information of motion. Therefore, the joints of both hands and feet in the representative group would not be included. In summary, a list of limb vectors is depicted as Table 2.

Table 2 List of limb vectors

No.	Description	No.	Description	No.	Description	No.	Description
1	Left elbow	3	Left thigh	5	Left arm	7	Left calf
2	Right elbow	4	Right thigh	6	Right arm	8	Right calf

3.3 Scoring Evaluation

After classifying the limb vectors, the angles among them are computed owing to the obtained data and the reference model as follow

$$\cos\phi = \frac{x_{ref}x_h^i + y_{ref}y_h^i + z_{ref}z_h^i}{\sqrt{x_{ref}^2 + y_{ref}^2 + z_{ref}^2}\sqrt{\left(x_h^i\right)^2 + \left(y_h^i\right)^2 + \left(z_h^i\right)^2}} \tag{1}$$

where,

$x_{ref}, y_{ref}, z_{ref}$: the coordinate of limb vector for reference model.

x_h^i, y_h^i, z_h^i: the coordinate of vector of hand-feet currently.

$\{\phi_1, \phi_2, \phi_3, \phi_4, \phi_5, \phi_6, \phi_7, \phi_8\}$: a set of relative angles for each vector in Table 2.

Equation (1) indicates the information of differences between the current gesture and the reference model. In reality, there are several tolerances for the inequivalence in vision. Henceforth, the weights are essential to insert for these compensations. In this study, the pseudo-distance Λ to specify total differences is recommended as below

$$\Lambda = [w_1(\phi_1 + \phi_2) + w_2(\phi_3 + \phi_4) + w_3(\phi_5 + \phi_6) + w_4(\phi_7 + \phi_8)] \tag{2}$$

where,

w_1, w_2, w_3, w_4: weights of elbow, arm, thigh and calf respectively.

To estimate these weights, the proposed system automatically collects a number of gestures in recent time. In the database, the rating level Tol fluctuates $\pm$ 10% and the average of these values. In detail, the weight w_i is

$$w_i = \frac{1}{Tol\left(\sum_{n=4}^{1} \frac{1}{Avg_n}\right)} \tag{3}$$

where,

Avg_n: the nth average value.

Thus, final score D could be evaluated as

$$D = \sigma_0\left[\left(\Lambda_{ref} - \Lambda\right) \times \frac{10 - D_{ref}}{D_{ref}}\right] \tag{4}$$

where,

D_{ref}, Λ_{ref}: the reference score and distance from user correspondingly.

σ_0: the tuning parameter depending on experiences.

4 Results of Study

Through the numerous evaluations of the player performance, the effectiveness and feasibility of the proposed approach should be assessed. The following software which is written by C/C++ programming language, integrates the deep learning technique, data analysis and the scoring mechanism. It could work well on personal computer with core i7 3200 MHz, OS Windows 10 Professional. This system requires two digital cameras to capture visual data from two sides. The first one which provide images, motion recognition and emotion detection, locates in front of player as direct view as Fig. 7. The second camera plays a role as supplement to enhance the accuracy of overall system. Especially, this system is also available for the indoor environment where lacks of light.

Generally, there are three levels of physical exercises for teenagers as Table 3. In the easy level, player should be familiar with activities such as clapping her hand or hand shaping. In the second level, both more actions and timing checks are required. Lastly, the difficulties are to ensure the pose, direction and speed of movement in respect to reference model in hard level.

In the second tab of software as Fig. 8, it is proper to train each action individually. Players must imitate and repeat until they could handle the exercise fluently. After

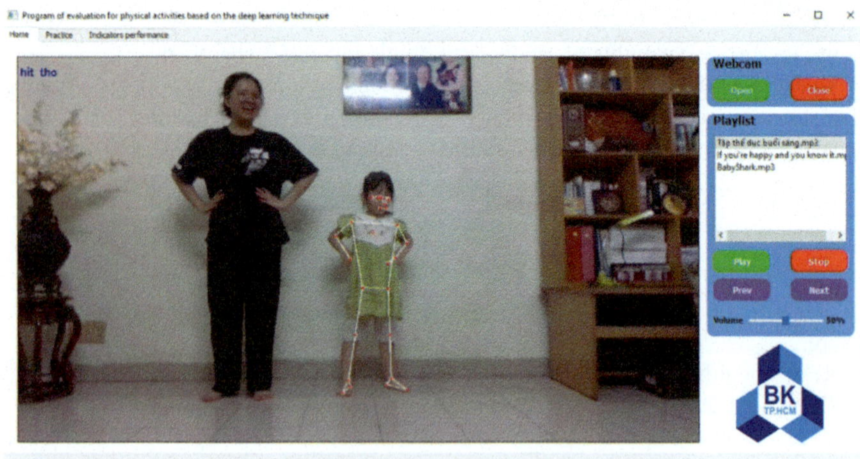

Fig. 7 User control panel and data capture

Table 3 List of exercise levels

No	Name	Level
1	Baby shark	Easy
2	Tập thể dục buổi sáng	Medium
3	If you're happy and you know it	Hard

every practice, the scoring mechanism would return a reward which is as premise to improve their performance.

Instantly, the process of data analysis is activated when the physical exercise completes. Whole assessment is briefly summarized and visually displayed. Some key factors to indicate the player performance are the timeline of actions or the average scores. In addition, the emotion of player is also estimated in order to measure how much their interest is. From those indicators as Fig. 9, this system could release a reasonable advice to encourage players in their performance.

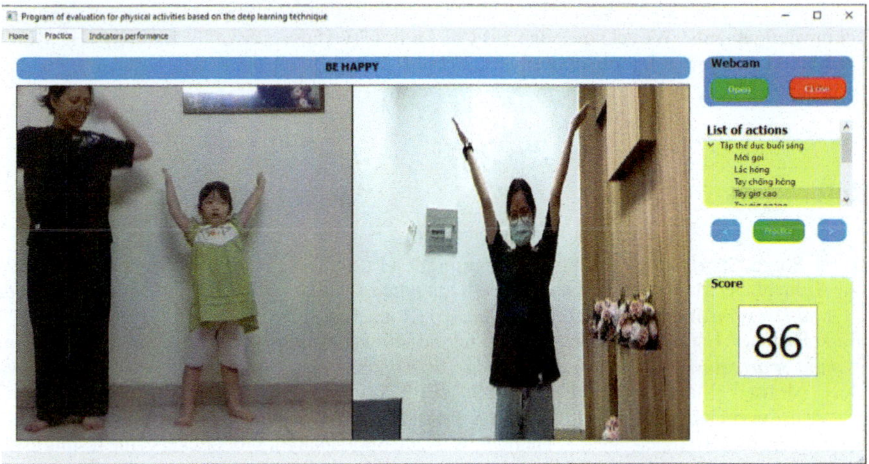

Fig. 8 Evaluation mode and the scoring mechanism

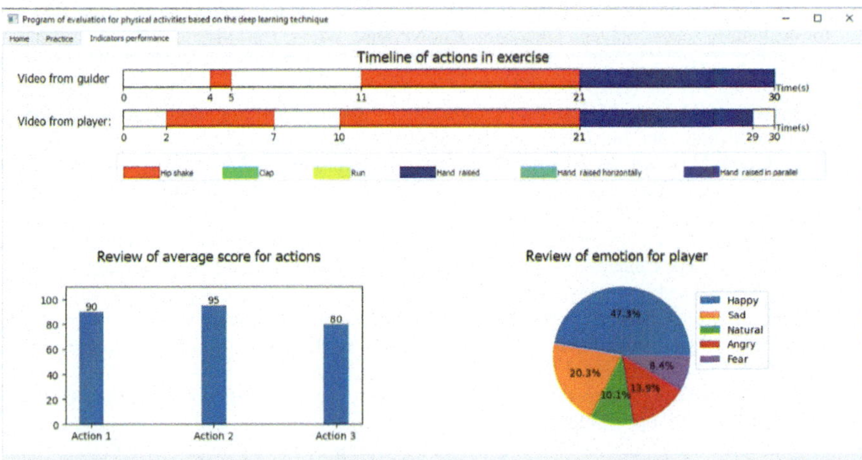

Fig. 9 Result of data analysis for the player performance

5 Conclusions

In this paper, a novel approach of vision-based application using the deep learning technique was presented. The hardware platform and the setup of experimental scenario were launched. Later, the scoring mechanism was evaluated by mathematical model. It could produce the assessment of player performance precisely and directly. Several indicators were also implemented in this system. From the successful implementation, the proposed solution has been validated to apply in education, sport analytics or gesture-based control.

Acknowledgements We acknowledge Ho Chi Minh City University of Technology (HCMUT), VNU-HCM for supporting this study.

References

1. N. Singh, Sport analytics: a review. Learning **9**, 11 (2020)
2. A. Jayal, A. McRobert, G. Oatley, P. O'Donoghue, *Sports Analytics: Analysis, Visualisation and Decision Making in Sports Performance* (Routledge, 2018)
3. V. Ratten, P. Usmanij, Statistical modelling and sport business analytics, in *Statistical Modelling and Sports Business Analytics* (Routledge, 2020), pp. 1–9
4. L. Morra, F. Manigrasso, F. Lamberti, SoccER: Computer graphics meets sports analytics for soccer event recognition. SoftwareX **12**, 100612 (2020)
5. K. Kapadia, H. Abdel-Jaber, F. Thabtah, W. Hadi, Sport analytics for cricket game results using machine learning: an experimental study. Appl. Comput. Inf. (ahead-of-print) (2020)
6. J. Fernández, L. Bornn, D. Cervone, Decomposing the immeasurable sport: a deep learning expected possession value framework for soccer, in *13th MIT Sloan Sports Analytics Conference* (2019
7. G. Liu, Y. Luo, O. Schulte, T. Kharrat, Deep soccer analytics: learning an action-value function for evaluating soccer players. Data Min. Knowl. Disc. **34**(5), 1531–1559 (2020)
8. V. Sarlis, C. Tjortjis, Sports analytics—evaluation of basketball players and team performance. Inf. Syst. **93**, 101562 (2020)
9. S. Wenninger, D. Link, M. Lames, Performance of machine learning models in application to beach volleyball data. Int. J. Comput. Sci. Sport **19**(1), 24–36 (2020)
10. D. Formenti, M. Duca, A. Trecroci, L. Ansaldi, L. Bonfanti, G. Alberti, P. Iodice, Perceptual vision training in non-sport-specific context: effect on performance skills and cognition in young females. Sci. Rep. **9**(1), 1–13 (2019)
11. K. Apostolou, C. Tjortjis, Sports Analytics algorithms for performance prediction, in *2019 10th International Conference on Information, Intelligence, Systems and Applications (IISA)* (IEEE, 2019), pp. 1–4
12. M. Mutz, J. Müller, A.K. Reimers, Use of digital media for home-based sports activities during the COVID-19 pandemic: results from the German SPOVID survey. Int. J. Environ. Res. Public Health **18**(9), 4409 (2021)
13. M. Nibali, The data game: analyzing our way to better sport performance, in *Sport Analytics*. (Routledge, 2016), pp. 71–97
14. L. Goebeler, W. Standaert, X. Xiao, Hybrid sport configurations: the intertwining of the physical and the digital (2021)
15. J. He, Y. Bai, Fuzzy analytic hierarchy process based volleyball quality evaluation for college teaching, in *2018 International Conference on Intelligent Transportation, Big Data & Smart City (ICITBS)* (IEEE, 2018), pp. 669–672

16. K. Peng, Y. Zhao, X. Sha, W. Ma, Y. Wang, W.J. Li, Accurate recognition of volleyball motion based on fusion of MEMS inertial measurement unit and video analytic, in *2018 IEEE 8th Annual International Conference on CYBER Technology in Automation, Control, and Intelligent Systems (CYBER)* (IEEE, 2018), pp. 440–444
17. Y. Yoon, H. Hwang, Y. Choi, M. Joo, H. Oh, I. Park, J.H. Hwang, Analyzing basketball movements and pass relationships using realtime object tracking techniques based on deep learning. IEEE Access **7**, 56564–56576 (2019)
18. A. Mujahid, M.J. Awan, A. Yasin, M.A. Mohammed, R. Damaševičius, R. Maskeliūnas, K.H. Abdulkareem, Real-time hand gesture recognition based on deep learning YOLOv3 model. Appl. Sci. **11**(9), 4164 (2021)
19. R. Aminuddin, M.A. Shamsudin, N.I.F.A. Wahab, Mobile application framework for monitoring target heart rate zone during physical exercise using deep learning, in *2021 IEEE 9th Conference on Systems, Process and Control (ICSPC 2021).* (IEEE, 2021), pp. 1–6
20. M. Ivasic-Kos, M. Kristo, M. Pobar, Person detection in thermal videos using YOLO, in *Proceedings of SAI Intelligent Systems Conference* (Springer, Cham, 2019), pp. 254–267
21. M. Krišto, M. Ivasic-Kos, M. Pobar, Thermal object detection in difficult weather conditions using YOLO. IEEE Access **8**, 125459–125476 (2020)

Machine Learning-Based Approaches for Internal Organs Detection on Medical Images

Duy Thuy Thi Nguyen, Mai Nguyen Lam Truc, Thu Bao Thi Nguyen, Phuc Huu Nguyen, Vy Nguyen Hoang Vo, Linh Thuy Thi Pham, and Hai Thanh Nguyen

Abstract In recent years, deep learning algorithms have risen in popularity and growth because they have achieved outstanding results in various disciplines, including face recognition, handwritten character identification, image classification, object detection, and object segmentation on images. These systems are based on computer self-learning algorithms that use deep learning. Furthermore, deep learning algorithms offer to open up a promising research direction for medical image analysis applications. Intelligent robots, integrated expert systems leveraged from machine learning-based advancements, are expected to support doctors performing complicated surgeries. This study has conducted image processing techniques in the identification of human internal major organs, including the heart, lung, trachea, and liver using various image segmentation algorithms. The work is expected to assist in the diagnosis process and automatic human parts detection for techniques of ultrasound examinations and surgery activities.

1 Introduction

In recent years, the strong development of computer science, especially machine learning and artificial intelligence (AI), has opened up many potential development directions in the medical field. For example, the rapid development of imaging techniques in medicine has created a huge amount of medical data.[1,2] Therefore, using AI to help people find useful information quickly is a necessary and impor-

D. T. T. Nguyen · M. N. Lam Truc · T. B. T. Nguyen · P. H. Nguyen · V. N. Hoang Vo · L. T. Thi Pham · H. T. Nguyen (✉)
Can Tho University, Can Tho, Vietnam
e-mail: nthai.cit@ctu.edu.vn

[1] https://dangcongsan.vn/khoa-hoc-va-cong-nghe-voi-su-nghiep-cong-nghiep-hoa-hien-dai-hoa-dat-nuoc/diem-nhan-khoa-hoc-va-cong-nghe/ung-dung-tri-tue-nhan-tao-trong-kham-chua-benh-566810.html.

[2] https://drbinh.com/tri-tue-nhan-tao-ai-loi-ich-va-nhung-dot-pha-trong-nganh-duoc-pham.

N. H. Phuong and V. Kreinovich (eds.), *Deep Learning and Other Soft Computing Techniques*, Studies in Computational Intelligence 1097,
https://doi.org/10.1007/978-3-031-29447-1_9

tant step to develop the medical profession as well as increase ability to successful treatment for patients.

Image segmentation tasks are important and difficult. Nowadays, image segmentation methods are developing to be faster and more accurate for emergency cases. By combining many new theories and technologies, researchers are finding a common segmentation algorithm that can be applied to this type of image.

With the advancement of medical treatment, all kinds of new medical imaging devices are becoming more popular. The types of medical imaging widely used in the clinic are mainly computed tomography (CT), magnetic resonance imaging (MRI), positron emission tomography (PET), X-ray, and Ultrasound imaging (UI). Doctors use CT and other medical images to assess a patient's condition. Therefore, research on medical image processing has become the focus of attention in computer vision. With the rapid development of artificial intelligence, especially deep learning, image segmentation methods based on deep learning have achieved good results in image segmentation. Compared with traditional machine learning and computer vision methods, deep learning has certain advantages regarding segmentation accuracy and speed. The use of deep learning to segment medical images can help doctors effectively confirm the size of the diseased tumor and quantitatively evaluate the effect before and after treatment, greatly reducing the doctor's workload [1].

The organ segmentation task is an essential step in data analysis for many areas of medical research. The research aims to identify parts of the human body to help make the diagnosis easier and more accurate and to be pre-research to support the development of surgical robots based on automatic techniques. The main contribution of the research is to evaluate the efficiency of various segmentation approaches (UNET, SegNet, and a Fully Connected Network (FCN)) on important human organs. First, we collected data with Liver images from,[3] X-ray images for lung segmentation,[4] and other sources from Kaggle. Then, we leverage segmentation algorithms such as UNET, SegNet, and a Fully Connected Network (FCN) for classifying and marking major organs in medical images, such as the heart, trachea, lung, and liver.

2 Related Work

Many units and organizations in Vietnam are currently involved in medical image processing research. For example, Vingroup Corporation has launched VinDr—the first version of a medical image analysis solution with comprehensive application of Artificial Intelligence (AI) technology, an applied research project of the Big Data Research Institute (VinBDI) [2]. Vingroup Corporation started testing at 108 Central Military Hospital, Hanoi Medical University Hospital, and Vinmec Times City International Hospital. A special feature of this solution is the application of artificial intelligence (AI) technology on the medical image storage and transmission (PACS)

[3] https://www.ircad.fr/.

[4] https://paperswithcode.com/dataset/montgomery-county-x-ray-set.

platform. In the first step, VinDr will support two functions: Diagnosis of lung diseases on chest X-ray images [3] and diagnosis of breast cancer on mammograms [4]. In addition, FPT Software company, with experts in artificial intelligence (AI) from the University of Toulouse (France) and professors and dermatologists in Vietnam, has researched and developed the application. DeepCinics dermatology diagnostics. This application uses technologies such as AI and machine learning (Machine Learning) to create a 4.0-trending medical examination and diagnosis system with a diagnostic accuracy of about 80–90%.[5]

A typical example of the applications of AI in surgery is that robots are gradually being used in laparoscopic surgery to perform surgeries such as ENT surgery, neurosurgery, gynecological surgery, and gastrointestinal surgery hepatobiliary system [5]. Among them, 4 outstanding robotic systems are being applied, including Da Vinci endoscopic surgery robot [6], Renaissance spinal surgery robot, MAKOplasty knee surgery robot [7] and ROSA neurosurgery robot [8]. Specifically, in 2012, Viet Duc Hospital was the first unit to deploy the Renaissance precision positioning robot application in spine surgery. At Bach Mai Hospital, in 2015, the hospital built 2 new operating rooms with 2 robotic surgical systems, MAKO and ROSA, the most modern, on par with other countries worldwide to serve and treat patients' diseases. After the very effective application of robots in spine surgery, in February 2017, Bach Mai Hospital (Hanoi) continued to put into use 2 surgical robots to treat knee pathology and neurosurgery.[6] Moreover, Cho Ray Hospital in Vietnam has officially deployed the Da Vinci robotic surgery system to treat many types of cancer such as prostate cancer, kidney cancer, bladder cancer, colorectal cancer, lung cancer, choledochal cystectomy, pyelonephritis-ureteral junction reconstruction, liver cancer.[7]

Nowadays, recognition techniques through deep learning models have been studied and applied more and more in the medical field, especially in leading countries. There are extremely useful application fields such as artificial intelligence in Radiotherapy for cancer treatment [9]; Automate the segmentation of agencies, determining ROI; Computer-Aided Diagnosis (CADx) [10]; Mass detection [11]; Assist in endoscopic procedures [12].

One of the medical AI applications to mention is the RapidAI software,[8] which helps doctors diagnose and treat more quickly and accurately, storing the past images processing of patients after performing health checks such as CT scans, MRI scans, etc. These images will be immediately sent to the hospital's PACS system or provided to specialists via mobile device or computer [13]. Then, the specialist will analyze it to decide on the treatment for the patient.

Softneta Company designed MedDream DICOM Viewer [14] for the web-based PACS server to provide a fast and reliable way to search, view, analyze and diag-

[5] https://vnexpress.net/dung-ai-chan-doan-benh-ve-da-3997315.html.

[6] http://medinet.gov.vn/cai-cach-hanh-chinh-y-te-thong-minh/gioi-thieu-san-pham-binh-chon-giai-thuong-y-te-thong-minh-cua-benh-vien-nhan-da-c4714-20030.aspx.

[7] https://kcb.vn/benh-vien-bach-mai-ung-dung-he-thong-robot-trong-phau-thuat-khop-goi-va-phau-thuat-than-kinh.html http://arxiv.org/abs/1704.06825.

[8] https://www.rapidai.com/about.

nose images, signals, and video files. Healthcare from anywhere and on any device, assisting medical professionals in diagnosis and treatment. The DICOM viewer was developed using a responsive design that allows medical imaging to be accessed on computers, tablets, smartphones, or other devices with mobile viewing capabilities. MedDream DICOM Viewer's 3D imaging simplifies the technique of reconstructing 3D images from slices of 2D images. In addition, the technology provides various views from the original data using 3D shaping techniques such as MPR and MIP.[9] In addition, the 3D Slicer application is a widely used software platform for analyzing and visualizing medical images and studying image-guided therapies. This open-source software is suitable for operating systems such as Linux, MacOSX, and Windows. The software includes features such as Multi-organism and support for many image formats such as MRI, CT, and US.

3 Methods

3.1 *Overview Structure*

Model structure: The U-net model consists of a contracting path with four encoding blocks, followed by an expanding path with four decoding blocks. Each cipher block consists of two consecutive 3×3 convolutional layers, followed by a max-pooling layer with a stride of 2 to perform downsampling. In contrast, the decoding blocks include a transpose-Conv layer to perform upsampling, concatenation with the corresponding feature map from the contracting path and two 3×3 convolutional layers. All Conv classes are followed by Batch normalization and activation = ReLU. In the final layer, a 1×1 convolution is used to map the output from the final decoder block to a feature map, where Sigmoid is applied to normalize the output image of the model to the same scale as the input. As shown in Fig. 1, the medical image processing can be explained as follows: Labels and training sets are used as input for the training model. For each epoch, when the epoch value is not maxed, it will output the output and the loss value (loss when performing training). Therefore, continuing to iterate through each epoch and, at the same time, saving the model is the best result after training. Finally, use the saved model to test with the corresponding test set and labels. The result obtained is the prediction of the model.

3.2 *Data Description*

Chest X-ray data was collected from 2 data sets:

---

[9] https://www.softneta.com/products/meddream-dicom-viewer/.

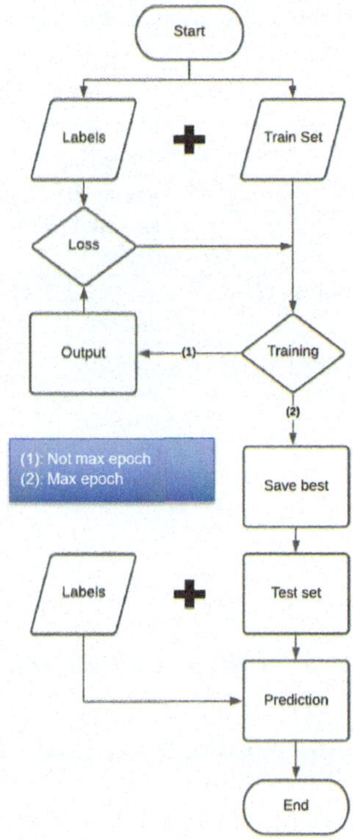

Fig. 1 Flow diagram for medical image processing

- Montgomery County X-ray Kit: The images in this dataset were obtained from the Tuberculosis Control Program of the Department of Health and Human Services of Montgomery County, MD, USA. This set includes 138 X-rays - posterior-anterior chest radiographs. All images are de-identified and available with left and right PA-view lung masks in png format.
- X-ray kit for Shenzhen hospital: The chest X-ray images in this dataset were collected and provided by Shenzhen No. 3 Hospital in Shenzhen, Guangdong Province, China. The images are in png format.

Liver CT data were obtained from the 3D IRCAD 3 dataset used to test and evaluate methods of segmentation and detection of intrahepatic tumors. The IRCAD dataset consists of 20 intravenous phase-enhanced CT volumes obtained using different CT scanners. Heart, lung, and tracheal CT image data was collected from Kaggle. This dataset includes 3D images and masks (.nrrd) that 3D Slicer has processed to produce a dataset that includes 2D slices that are png images. The description of the collected data is shown in Table 1.

Table 1 Description of data collected

Part	Dataset name	Format	Amount	Size	Resize for training
Lung	Montgomery dataset	jpg	138 CXR image	4020 × 4892	512 × 512
	Shenzhen Hospital (SH) dataset	jpg	662 images	4020 × 4892	512 × 512
Liver	3D IRCAD 3 dataset	.nii	20	512 × 512	
Lung-Heart-Trachea	OSIC Pulmonary Fibrosis Competition from Kaggle	.nrrd	87		
		.png	17012	512 × 512	

3.3 *Image Processing*

3.3.1 Lung X-ray Image Processing

In order to match the designed U-Net network architecture, preprocessing will be performed for all data. The images of the original dataset are of different sizes for each image. Therefore, we resized these images and their respective masks to 512 × 512 pixels to train the model. Furthermore, enhanced data generation is performed on the images and their respective masks to create variety and assess model accuracy.

To perform lung segmentation via chest X-ray, we use a segmentation technique that separates the image into different regions (presumptive lung and non-lung areas). Each of these regions comprises a set of pixels that share certain common characteristics. This image-processing technique simplifies an image's representation into something more useful and easier to use. U-Net is used to segment images based on ground truth images marked as lungs. U-Net performs well when the prediction results are relatively accurate for both left and right lungs based on extracting features and reconstructing the predicted image, such as the input image size. Moreover, the use of X-ray images also helps to increase the accuracy of the model when predicting because the lungs have strong X-ray transmission, so the gray value in the corresponding region of the CXR image will be low. In contrast, the surrounding tissues and organs have a lower ability to transmit X-rays, so the corresponding gray value of CXR will also be higher. The higher the gray value, the more blurred the image is, and the harder it is to perceive in Fig. 2.

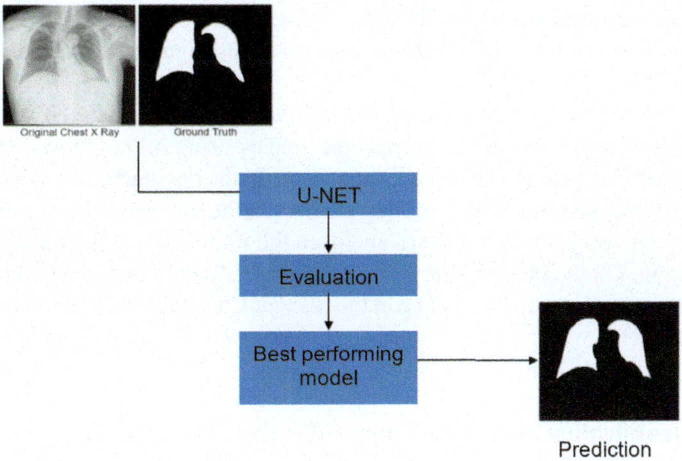

Fig. 2 Diagram of chest X-ray image processing

3.3.2 Liver CT Image Processing

For liver CT images, because it is a 3D images with nii file format, it is necessary to convert the CT images into a format that the training model can understand. Therefore, we have converted the data set from nibabel (.nii) to NumPy(.npy) in this study. To implement this method, we processed the data in the following way: load the 3D data and cut it vertically (in the z-axis), then collect the slice images containing the liver (2D format) and store it as data numpy.

For image preprocessing, we truncated the image intensity values of the CT scans within a fixed range [0, 200] to remove extraneous details and enhance the contrast of the image. As a result, as illustrated from [15] in Fig. 3, the boundaries of the liver were clearer after the image mentioned above preprocessing.

For liver segmentation, we use the rough-to-fine method to reduce the calculation time and improve the segmentation's accuracy. A 2D slice-based U-grid was used to determine the proximal location of the liver. At the training stage, we performed double normalization for the 3D CT image I and the corresponding base segment Y. On the one hand, we divided each I and Y into a set of 2D slices and resized them. Then, size them to a fixed 256 × 256 image size. These normalized 2D images and

Fig. 3 Slices of liver CT images before and after intensity resection with a fixed range [0, 200]

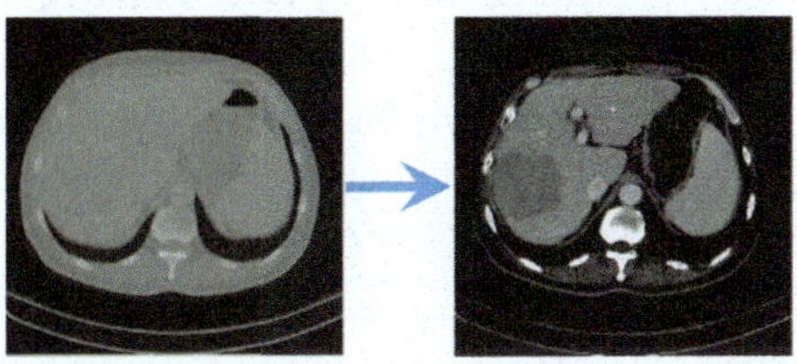

I and Y pairs are then fed into a 2D U-Net to fit the raw segmentation model, which focuses on learning the distinguishing features to distinguish liver and background discrimination.

On the other hand, we resize each 3D volume (I and Y) to a fixed voxel size of $1 \times 1 \times 1$ mm^3 and crop these normalized images using a liver mask. The liver is segmented from coarse to fine at the testing stage. In the beginning, we also normalize the image twice, just like in the training phase. On the one hand, we cut and resized the 3D CT test images into 2D slices and used the trained U-net to predict the rough liver segment. On the other hand, we resize both the 3D CT image and the raw liver segment to normal voxel size and crop the resized CT image around the resized raw liver segment.

3.3.3 Identification of Heart—Lungs—Trachea Through CT

This dataset was used from the lung segmentation dataset.[10] This data originally consisted of nrrd files that were saved in tensor format with masks corresponding to the labels: (lung, heart, trachea) as small arrays using a pickle. Each tensor has the following shape: several slices, width, height, and several layers, where the width and height of the slice are the individual parameters of each tensor id and the number of layers = 3. In addition, the data is saved as an RGB image, where each image corresponds to an ID slice, and their mask image has channels corresponding to three layers: (lung, heart, and trachea). This data can be depicted as Fig. 4.

By dividing the data into 3 layers: heart, lung, and trachea for input and mask images, respectively, as mentioned, the model will perform identification on each RGB image—which is a slice by the ID of the original CT image. The information of the image will be divided into regions of the heart, lungs, trachea, and areas that do not belong to those three parts. After recognizing each slice and getting the prediction results, we will combine the slices by this ID into a video. After processing the image, the colors and parts will be based on the corresponding mask in our dataset, namely the lung area corresponding to the blue mask, the heart area will be red, and the air area will be red. The stem will be green.

4 Experimental Results

4.1 Divide Training and Testing Set

The image data set is divided by hold-out into training and testing sets, with ratios of 0.9 and 0.1. The training model compiled using the optimizer is ADAM. The input image is resized to 256×256 to reduce training time and also memory requirements.

[10] https://kaggle.com/sandorkonya/ct-lung-heart-trachea-segmentation.

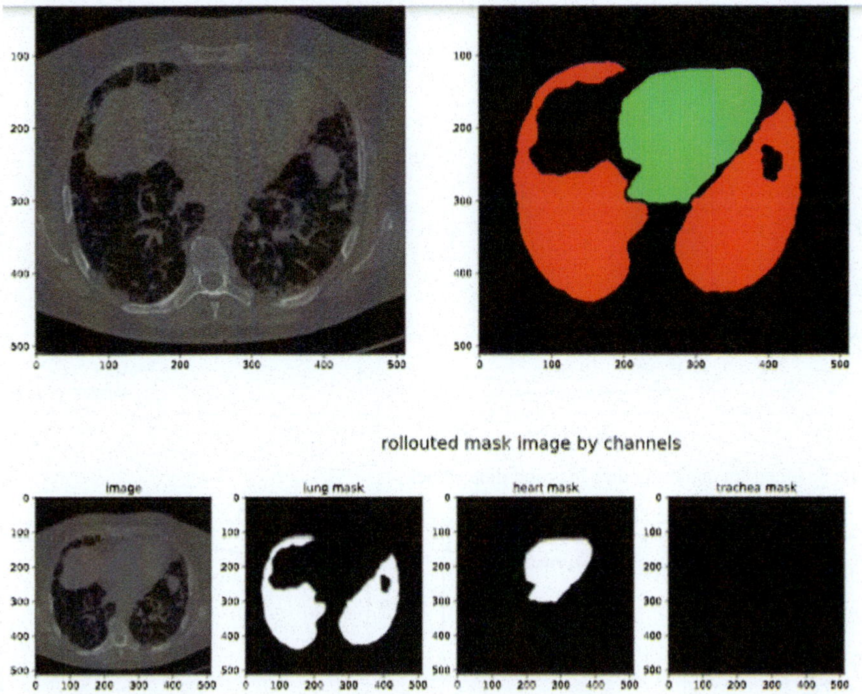

Fig. 4 Example of a dataset used for heart-lung-tracheal recognition

4.2 Experimental Results on Organs Segmentation

4.2.1 Lung

The study used the evaluation method based on pixel accuracy. That is, accuracy will be calculated as the percentage of pixels in the image that is correctly classified. Figure 5 depicts the loss value and accuracy of the proposed model. The results show that accuracy (accuracy), accuracy validation (accuracy) and, loss value (loss), loss validation (validation loss value) were obtained during the training process of 50 epochs. It can be seen that the accuracy and validation accuracy are almost over 95%, which means that the proposed model can achieve good performance. Furthermore, the accuracy increases and the loss decreases while the trained model produced the best model that was saved after training to the last epoch.

4.2.2 Liver

Evaluate the model through the Dice coefficient value used for liver segmentation. The dice coefficient is often used to quantify the performance of image segmentation

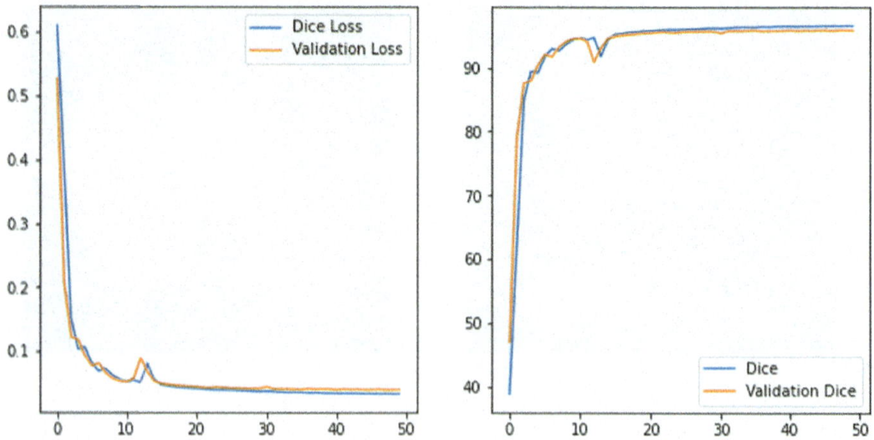

Fig. 5 Diagram depicting dice loss and dice values

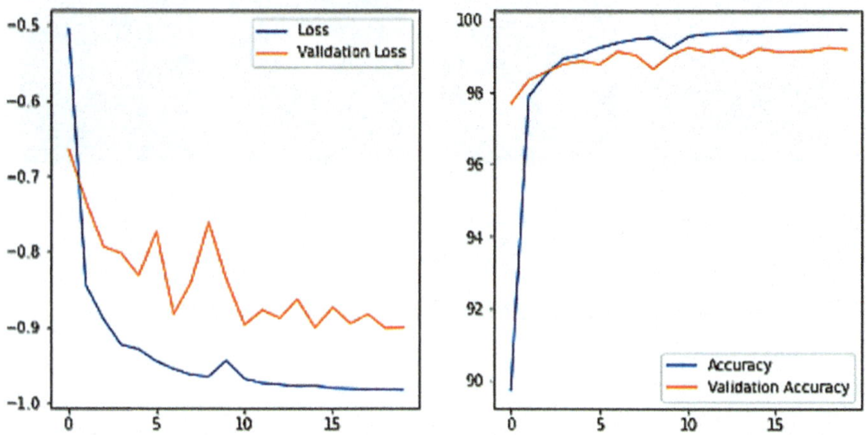

Fig. 6 Loss and dice values during epochs on Liver segmentation

methods. The dice coefficient is an evaluation method based on the similarity ratio of objects. It is the ratio of 2 times the overlapping area (the overlap) of the two images divided by the sum of the pixels of the two images, calculated as follows (Fig. 6):

$Dice = \frac{2\,|X \cap Y|}{|X|+|Y|}$, X, Y is each image's total number of pixels.

For binary data, Dice coefficient $Dice = \frac{2TP}{2TP+FP+FN}$, where TP, FP, TN, and FN are True Positive, False Positive, True Negative, and False Negative, respectively.

Figure 7 depicts the Dice value of the Train set and Test set while training model with 20 epochs. It can be seen that the Dice value is generally increasing in both data sets. The same ratio between labels and output when training the model increases over each epoch. This proves that the proposed model can achieve good performance.

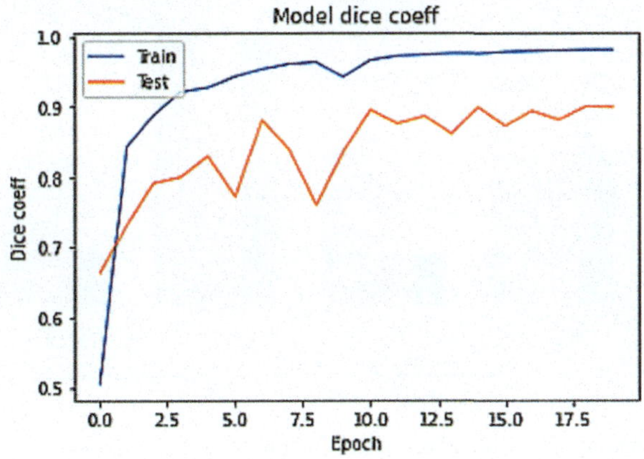

Fig. 7 The performance in Dice value

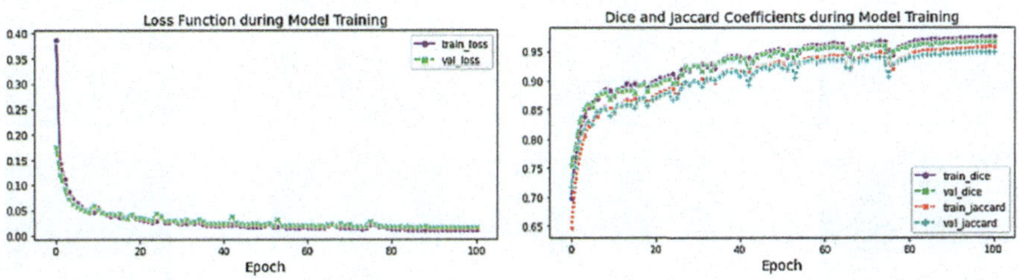

Fig. 8 Diagram describing the results

4.2.3 Heart, Lungs, Trachea

Figure 8 depicts the accuracy and the loss value results when training the model by two evaluation methods, the Jaccard index and the Dice coefficient. The results show that in both methods, the loss value decreases, and the accuracy increase over 100 epochs. From that, it can be seen that the proposed model can achieve good performance for multipart segmentation.

4.3 Results Comparison with Various Segmentation Algorithms

Figure 9 shows the Accuracy, Recall (R), Specificity (SP), and Jaccard ratios of three models commonly used in segmentation problems: FCN (Fully Connected Network), SegNet, and U-Net.

From the comparison results in Fig. 9, it can be seen that the indexes of FCN are lower than those of other algorithms. For FCN, insufficient details in the decoding

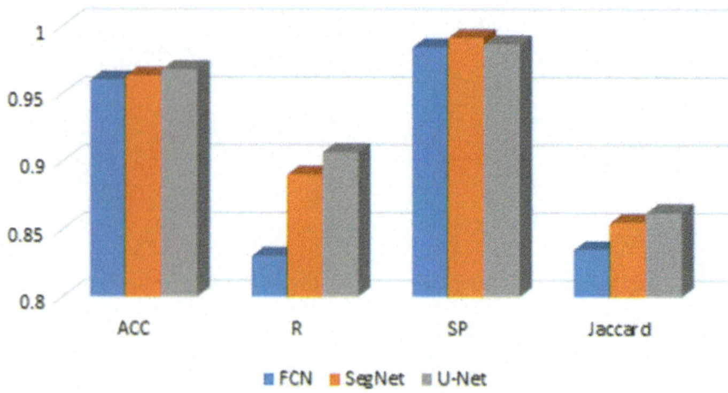

Fig. 9 Performance comparison on lung segmentation

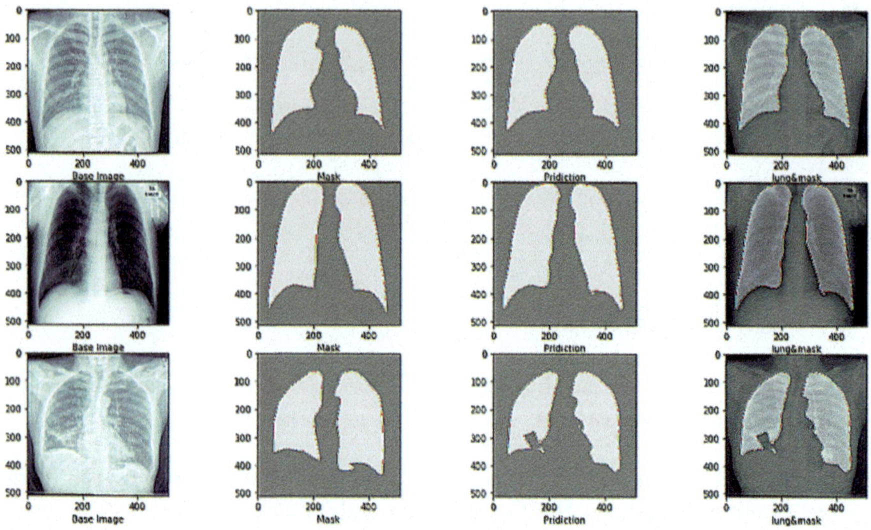

Fig. 10 Results obtained after performing the Test on the model

process may lead to poor segmentation results. Unlike FCN, SegNet records the max-pooling value and performs upsampling more accurately. This makes SegNet's segmentation performance better than FCN's. U-Net has better overall statistical results than FCN and SegNet. U-Net has Accuracy, Recall, and Jaccard as the most valuable indicators compared to the other two methods. These indicate that the segmentation efficiency of U-Net compared to other algorithms is better.

The results after performing the Test based on the trained model are shown in Fig. 10. The base image is the original image without segmentation, and the mask is the image's label. Prediction is the estimated result after performing testing. Finally, the prediction is combined with the base image.

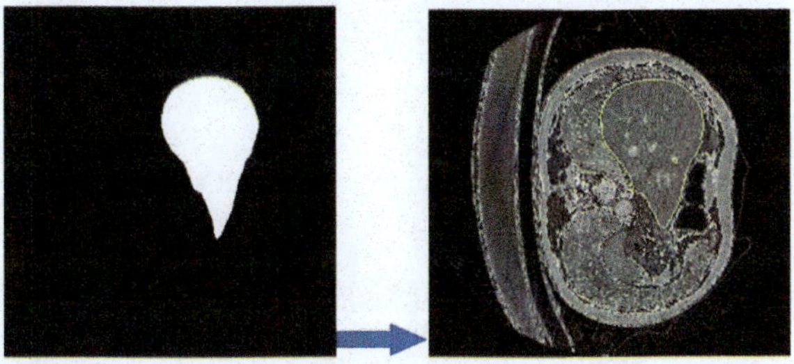

Fig. 11 The image on the left is the marked liver, and the right is the image of the liver corresponding to the marked liver

As for the liver, the original image of the data will not be presented here because it uses the input image in the .nii file format. Instead, the prediction results are shown through the label and prediction of the trained model. Because the prediction result presented as a video is a composite of the predicted image slices from the original input. Figure 11 is the result of several slices extracted from the prediction result video, so there may be a slight deviation between the actual results of the model and the representative image described above.

Figure 12 is a CT image of the processed video from the Heart-Lung-Trachea cut. The first image at the left root is the original CT image, and the middle and right images are 2 processed images. In this image, there are 3 colors corresponding to the 3 significant parts, namely the blue part. Therefore, positive will represent the lungs, light red will correspond to the heart, and the trachea will correspond to green.

5 Conclusion

In this study, we use the U-Net network to identify separate organs lung, liver, and heart. The results show that the recognition model has relatively high accuracy. This is the basis for further developing studies in identifying diseases and abnormalities in these organs using machine learning techniques. The results of the study contribute to the study of the use of machine learning in medical diagnostics. They can be used to simplify the diagnostic process and improve the management of the disease. While diagnoses have traditionally been validated by a single physician, allowing for the possibility of error, machine learning approaches can be seen as a breakthrough in improving diagnostic performance. In this case, the artificial intelligence system provides a diagnosis based on X-ray/CT images, which can then be confirmed by a physician, greatly reducing human and machine error. Therefore, this method can improve the diagnosis compared to traditional methods, improving the quality of treatment.

In the future, we will develop a system to detect diseases or abnormalities in internal organs on recognized medical images. Separating lesions into different categories, for example, by their size and treatment with different classifiers, can simplify

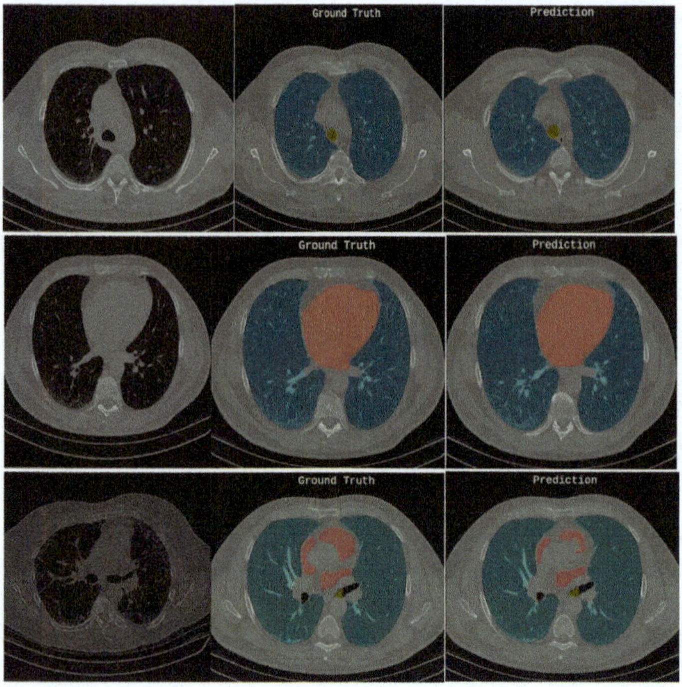

Fig. 12 Chest CT image before and after treatment of lung, heart, trachea

the task for all learners and help limit the problem. At the same time, they are developing models from 2D to 3D to be able to see the available organs and parts of the human body.

Acknowledgements This study is funded in part by Can Tho University, Code: THS2020-60.

References

1. Y. Fu, Y. Lei, T. Wang, W.J. Curran, T. Liu, X. Yang, A review of deep learning based methods for medical image multi-organ segmentation. Phys. Med. **85**, 107–122 (2021). https://doi.org/10.1016/j.ejmp.2021.05.003
2. H.Q. Nguyen, K. Lam, L.T. Le, H.H. Pham, D.Q. Tran, D.B. Nguyen, D.D. Le, C.M. Pham, H.T. Tong, D.H. Dinh, et al., VinDr-CXR: an open dataset of chest x-rays with radiologist's annotations. Sci. Data **9**(1), 1–7 (2022)
3. N.H. Nguyen, H.Q. Nguyen, N.T. Nguyen, T.V. Nguyen, H.H. Pham, T.N.M. Nguyen, A clinical validation of vinDr-CXR, an AI system for detecting abnormal chest radiographs (2021). arXiv:2104.02256
4. H.T., Nguyen, H.Q. Nguyen, H.H. Pham, K. Lam, L.T. Le, M. Dao, V. Vu, VinDr-Mammo: a large-scale benchmark dataset for computer-aided diagnosis in full-field digital mammography (2022). https://arxiv.org/abs/2203.11205

5. S.R., Wu, H.Y. Chang, F.T. Su, H.C. Liao, W. Tseng, C.C. Liao, F. Lai, F.M. Hsu, F. Xiao, Deep learning based segmentation of various brain lesions for radiosurgery (2020). https://arxiv.org/abs/2007.11784
6. C. Freschi, V. Ferrari, F. Melfi, M. Ferrari, F. Mosca, A. Cuschieri, Technical review of the da Vinci surgical telemanipulator. Int. J. Med. Robot. Comput. Assist. Surg. **9**(4), 396–406 (2013)
7. C. Batailler, A. Fernandez, J. Swan, E. Servien, F.S. Haddad, F. Catani, S. Lustig, Mako CT-based robotic arm-assisted system is a reliable procedure for total knee arthroplasty: a systematic review. Knee Surg. Sports Traumatol. Arthrosc. **29**(11), 3585–3598 (2021)
8. M. Lefranc, J. Peltier, Evaluation of the rosa^{TM} spine robot for minimally invasive surgical procedures. Expert. Rev. Med. Devices **13**(10), 899–906 (2016)
9. M. Santoro, S. Strolin, G. Paolani, G.D. Gala, A. Bartoloni, C. Giacometti, I. Ammendolia, A.G. Morganti, L. Strigari, Recent applications of artificial intelligence in radiotherapy: where we are and beyond. Appl. Sci. **12**(7), 3223 (2022). https://doi.org/10.3390/app12073223
10. K. Doi, Computer-aided diagnosis in medical imaging: historical review, current status and future potential. Comput. Med. Imaging Graph. **31**(4–5), 198–211 (2007)
11. R. Agarwal, O. Díaz, M.H. Yap, X. Lladó, R. Martí, Deep learning for mass detection in full field digital mammograms. Comput. Biol. Med. **121**, 103774 (2020)
12. L. Tanzi, P. Piazzolla, F. Porpiglia, E. Vezzetti, Real-time deep learning semantic segmentation during intra-operative surgery for 3d augmented reality assistance. Int. J. Comput. Assist. Radiol. Surg. **16**(9), 1435–1445 (2021)
13. Hamilton-Basich, M.: Rapidai awarded for ai-powered stroke imaging, diagnosis technologies. AXIS Imaging News (2021)
14. D. Haak, C.E. Page, K. Kabino, T.M. Deserno, Evaluation of DICOM viewer software for workflow integration in clinical trials, in *Medical Imaging 2015: PACS and Imaging Informatics: Next Generation and Innovations*, vol. 9418 (SPIE, 2015), pp. 143–151
15. Y. Zhang, B. Jiang, J. Wu, D. Ji, Y. Liu, Y. Chen, E.X. Wu, X. Tang, Deep learning initialized and gradient enhanced level-set based segmentation for liver tumor from CT images. IEEE Access **8**, 76056–76068 (2020)

Smart Bra Based on Impact and Acceleration Sensors Integrated Communication Techniques for Sexual Harassment Prevention

Linh Thuy Thi Pham, Thinh Phuc Nguyen, Khoi Vinh Lieu, Huynh Nhu Tran, and Hai Thanh Nguyen

Abstract The trauma of sexual assault can leave physical, emotional, and psychological wounds. Corrupt practices are still present in our society. This study introduces a smart system, namely 'SEcurity GIrl' (SEGI), which is integrated into a woman's bra. The device uses impact and acceleration sensors to recognize sexual harassment behaviors and send warnings via communication techniques to the victim's relatives for support. It can assist in alerting relatives about the dangers women and girls face, including when they are in active or passive situations. Notifications include calling and texting their Global Positioning System (GPS) location with 98% accuracy. Moreover, relatives can determine a new location if the attacker deliberately pulls the victim to another location. Furthermore, the device can generate a virtual fence to help parents receive warnings for their girls within a radius of 500 m. The system is deployed with low-cost, easy-to-use, simple message syntax and is an innovative solution for women's safety, especially teenagers.

1 Introduction

According to the United Nations Committee on the Elimination of All Forms of Discrimination against Women (CEDAW),[1] they stated that "Sexual harassment

[1] https://www.un.org/womenwatch/daw/cedaw/.

L. T. Thi Pham (✉)
Can Tho University of Technology, Can Tho, Vietnam
e-mail: pttlinh42@gmail.com

K. V. Lieu
Greenwich of University, Can Tho, Vietnam

T. P. Nguyen · H. Nhu Tran
Phan Van Tri high school, Can Tho, Vietnam

H. T. Nguyen (✉)
Can Tho University, Can Tho, Vietnam
e-mail: nthai.cit@ctu.edu.vn

N. H. Phuong and V. Kreinovich (eds.), *Deep Learning and Other Soft Computing Techniques*, Studies in Computational Intelligence 1097,
https://doi.org/10.1007/978-3-031-29447-1_10

includes unwelcome, sexual behavior such as physical contact, comments with a sexual meaning, the display or display of sexual images, and sexually explicit items; verbal or physical sexual solicitation". In addition, such conduct may offend honor and affect health and safety.[2] Currently, sexual harassment is at a high rate worldwide and in Vietnam, with remarkable statistics. In the US, 31% of female workers and 7% of male workers said they had been sexually harassed. As reported in,[3] there are up to 40–50% of women in Europe are harassed. This rate is also noticeable in Italy, where 55.4% of women aged 14–49 have experienced this condition. In particular, 30% of countries worldwide do not have laws prohibiting sexual harassment in the workplace, causing nearly 235 million female workers to have been at risk of becoming victims.[4] In the report in,[5] in the first 6 months of 2019 in Vietnam, the number of abused children increased dramatically, with 1,400 children, on average, 7 children are abused every day in the whole country. The subject of child abuse is very diverse as having different qualifications, ages, and occupations, including strangers and people who know the child, have relatives in the family; teachers and staff of educational institutions, officials and civil servants; pensioners, and the elderly. More than 95% of child abusers are men.

Several smart devices to protect women have been born. Each system uses a different technique to detect female insecurities. Among them is the study of GPS positioning to assist in knowing the location of the abused person. Or use the panic sensor to know the change in heart rate and body temperature. However, some devices are too bulky and unsightly, and the electrical capacity is not used enough for a working day, which should cause anxiety to users.

Our study presents a support device for women and children called "SEGI" to protect women from sexual abuse. The device is designed like a regular bra. When the abuser squeezes the bra or pulls on the bra strap, causing it to break, or if there is an action to push the victim, the device will immediately make a phone call and send the location to the phone number of a relative. In addition, relatives can turn on virtual fences with a radius of 500 m with simple message syntax. If they go out of the specified radius, the device will notify the family, helping to prevent a bad situation from happening in the future. At the same time, relatives can easily delete/change phone numbers and find the device via text message. The setting of the alarm signal can be modified with the number that the user can choose and fill in. For example, the user can fill in the number of relatives or police stations in the case the police officers allow. With this research, we hope to detect abuses promptly, protect victims and bring perpetrators to justice.

The rest of this study is organized as follows. Section 2 introduces the main related works. Section 3 presents a brief introduction to the technical requirements, prod-

---

[2] https://vbcwe.com/tin-tuc/quay-roi-tinh-duc-noi-lam-viec-va-nhung-con-so-biet-noi/31.

[3] https://www.eeoc.gov/statistics/charges-alleging-sex-based-harassment-charges-filed-eeoc-fy-2010-fy-2021.

[4] https://www.ilo.org/global/topics/violence-harassment.

[5] https://cand.com.vn/Su-kien-Binh-luan-thoi-su/Moi-ngay-can-nuoc-co-7-tre-em-bi-xam-hai-i567171/.

uct design, construction of wiring diagrams, algorithmic flowcharts, and operating principles of SEGI. Section 4 exhibits and analyzes the obtained results. Finally, we conclude the study and discuss future work in Sect. 5.

2 Related Work

A smart security solution called a smart wearable device system is implemented using the Raspberry Pi3 for enhancing the safety and security of women/children [1]. It works as an alert as well as a security system. It provides a buzzer alert to the people near the user (wearing the smart device). The system uses Global Positioning System (GPS) to locate the user. It sends the user's location through SMS to the emergency contact and police using the Global System for Mobile Communications (GSM)/General Radio Packet Service (GPRS) technology. The device also captures the image of the assault and surroundings of the user or victim using a USB Web Camera interfaced with the device and sends it as an E-mail alert to the emergency contact soon after the user presses the panic button on the Smart wearable device system.

Smart ring (SMARISA) in [2] included Raspberry Pi Zero, Raspberry Pi camera, buzzer, and button to activate the services. This device was extremely portable and could be activated by the victim who was assaulted by clicking a button that would fetch her current location and capture the attacker's image via the Raspberry Pi camera. The location and the image's link will be sent to predefined emergency contact numbers or police via the victim's smartphone, thus preventing additional hardware devices/modules and making the device compact. The problem with this device is that no GSM module was not considered, and costly to develop such devices. Moreover, "Raspberry Pi", a mini-computer, is also power-hungry. Those devices might not work correctly because of the integration of various sensors in one module. The sensors may generate wrong readings in any situation that can activate the device. The design of those devices is too bulky to be carried around.

A smart device for women's safety was introduced in [3] with the emergency alert system by using pressure, pulse rate, and temperature sensors to detect a possible atrocity using outlier detection is proposed. This system detects and sends alerts for the dear ones with the location coordinates of the women without requiring her interaction in critical times. Moreover, it sends an emergency message automatically to relatives and nearby police stations.

Recent research in [4] said that for every 29 min, a woman got raped in India. The study's proposed idea is very effective as it automatically gets activated with a single pull of the bag's handle or by pressing the button. This work suggested a digital tote bag that will help them in an emergency. It consists of a GPS module, GSM module, and camera module connected with a Raspberry Pi board and Arduino. So women in trouble can use this device and even harm the attacker and escape from them.

The authors in [5] presented devices customized to learn the individual pattern of temperature and heartbeat and then find out the threshold for generating an alarm.

Thus, this paper deals with designing a wearable women's safety device that automatically reads and creates patterns such as body temperature and pulse rate during running. If readings are higher than the normal readings, then it will automatically call and message more than one person along with the location so that actions can be taken. We used temperature and pulse sensors to detect the woman's activity. The sensors' data will be sent to the cloud, where a machine learning algorithm (logistic regression) is applied to analyze the generated data. Sensors first collect the data in non-danger conditions to train the algorithm. After that, data is used for testing to gauge the accuracy and how close it is to our trained data; the more accurate the accuracy, the more surety of danger, and the emergency alarm will be on emergency contacts. Thirdly, this paper deals with scenarios where there is no internet facility. We used the ZigBee mesh network to overcome the internet problem, which helped the device send the data to multiple hop distances. Kabir et al. in [6] proposed an application-based wearable device. The primary function of this device is to send SMS and the victimâŁ™s current area to the closest police headquarters and family members. The application interface is designed so that the map indicates a safe location to survive a criminal attack. However, this device is not user-friendly for rural women. Many girls from rural areas are not familiar with mobile applications or may not have a smartphone. However, the form factor of this device is too large to carry easily. Another study in [7] also proposed a smart mobile application, BONITAA, which also warped with various features such as SMS and location sending via GSM, health support, medical support, counseling, and self-defense tips for the rape victims to avoid rape. Moreover, to acknowledge the problem of rural women, they integrated the "Bangla" language into their application and tried to make it user-friendly. However, the problem is that women unfamiliar with mobile applications may not relish the facilities of those applications. The author in [8] proposed a wearable device to ensure womenâŁ™s safety to avoid sexual assault. The method also introduced a mobile application. They designed their device using GSM, GPS (Global Positioning System), and the Wi-Fi module integrated with a Microcontroller. The device can also make the call and send the location to the pre-recorded numbers or the nearest police stations to avoid unlawful activities. The main problem is that the device needs the always-on internet for web server access. They spent a lot on building such a device with a mobile application. However, it can not be affordable for every woman in our society, and also, the application-based interface is not flexible for all end users.

To acknowledge these problems, we have also built our safety device to support women and children. The device is tiny and can be carried out efficiently daily. Moreover, our device is very cost-efficient, and people of all levels can afford it at a reasonable price.

3 Methods

3.1 Requirements

The device does not interfere with usage habits, is not too bulky to avoid loss of user aesthetics, and has a small weight (the lighter, the better). The total weight should be at most 1 kg, helping the user to use it for a long time. In addition, each battery charge for about 5 h allows for at least 1 day of use before having to charge the battery if, on average, they use 8 h per day to go to work. If the abuser intentionally touches the chest area or knocks over to commit an act of abuse, the device must automatically send an alert to the relatives via calls and messages with GPS location. Anti-intrusion detection accuracy must reach 90% or more. The average time must be 0.3 s or less. The product can distinguish between a fall or an attack by others to avoid calling too many times about relatives. The relative can determine the new location if the victim has been dragged to another location. When they encounter an unexpected problem, they can contact their relatives urgently through default distress calls and messages with just a simple tap, without having to have a phone attached. Moreover, if a relative needs to change their phone number, they can easily change it through the text message syntax. The price is as low as possible to suit the economic conditions of most user in Vietnam. The device can be used after 10 min of instruction. All women can use it easily.

3.2 Design for Women's Safety

The hardware design of SEGI includes 1 padded bra for women, skin color, size 32 in shown Fig. 1. The circuit board design inside the shirt is located on the right side of the chest to avoid affecting the heart, including Arduino Nano board, 2 Flex sensors, MPU 6050 tilt sensor, sim 800L, GPS, vibration motor, 2 Lipo batteries 3.7 V–1000 mAh, push button, power switch. The total weight of the shirt after the device is attached is 280 g.

Figure 2 shows the circuit diagram design of the system, including the Adruino Nano board acting as a central processing unit, receiving signals from the SIM 800L to process the signal, output the signal (Digital, PWM) to devices that implement vibration motor circuits, horns, speakers, push buttons). Power supply for the device to operate includes 2 lipo batteries 3.7 V–1000 mAh. Bluetooth HC 05 sends a signal to Arduino in a HIGH and LOW state. The vibration motor circuit receives the Bluetooth signal from the software on the phone to turn the massage feature on and off. FLEX bending sensor: when the sensor is bent, the sensor value will change. It will send an ARDUINO signal to encode the signal and send it to SIM 800L AND GPS to send positioning and distress signals to people. SIM 800L module reads the status of the emergency button. If the button is turned on, the SIM 800L will send messages and calls to the pre-set relatives. GPS locates user coordinates. Switch to

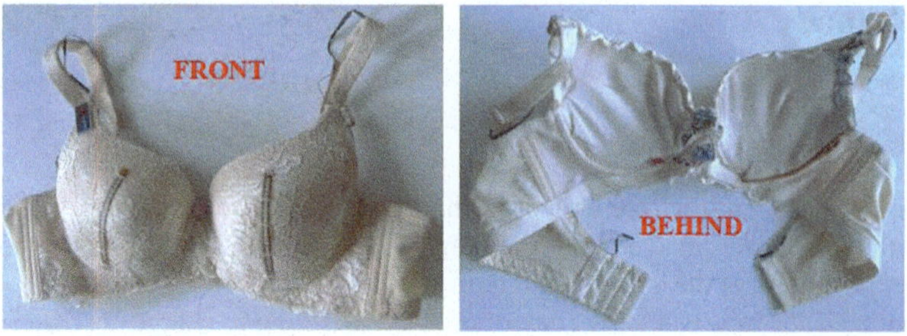

Fig. 1 Product design safety woman

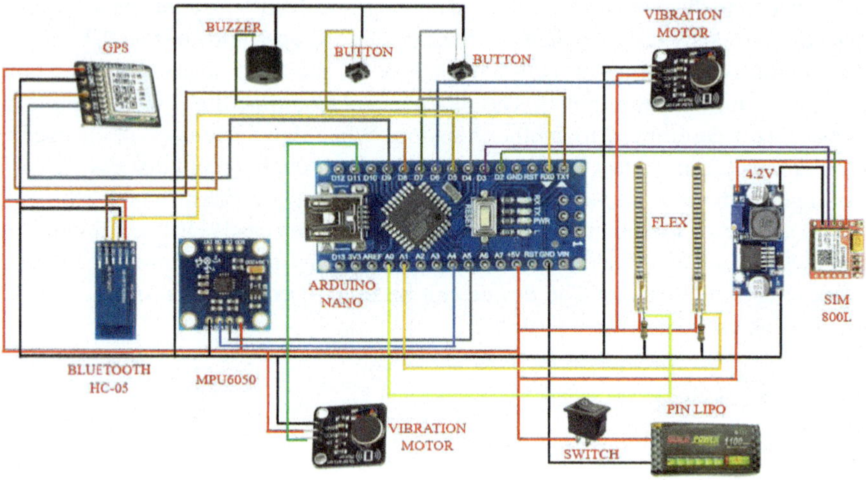

Fig. 2 Circuit diagram of segi

turn the device on and off. The tilt angle sensor MPU 6050 relies on the tilt relative to the axis to warn when necessary and send a value to the SIM 800L to send a distress signal to relatives.

The proposed algorithm flowchart is shown in Fig. 3. When the switch is on, the Arduino Module will be initialized, the data inputs will be activated, and the Flex sensor inputs will continuously check if the analog signal is less than 420 (it means someone acts on the product to warp the sensor). If yes, it will trigger a phone call to notify relatives and send GPS coordinates there. The 800L SIM module will check if the push button is activated (by the user in danger of pressing it). If yes, it will call and send a message of GPS coordinates to the pre-set phone number. Sim800L received the #RS# syntax will delete the previously saved phone number. If wrong, check the #ADMIN# syntax will add a new relative's phone number. If wrong, check the #ADR# to turn on the horn and send GPS coordinates to relatives' phones. If the #ON# syntax is received to turn on the virtual fence, the #OFF# syntax will turn off the virtual fence. Default virtual fence with a radius of 500 m. The tilt angle sensor

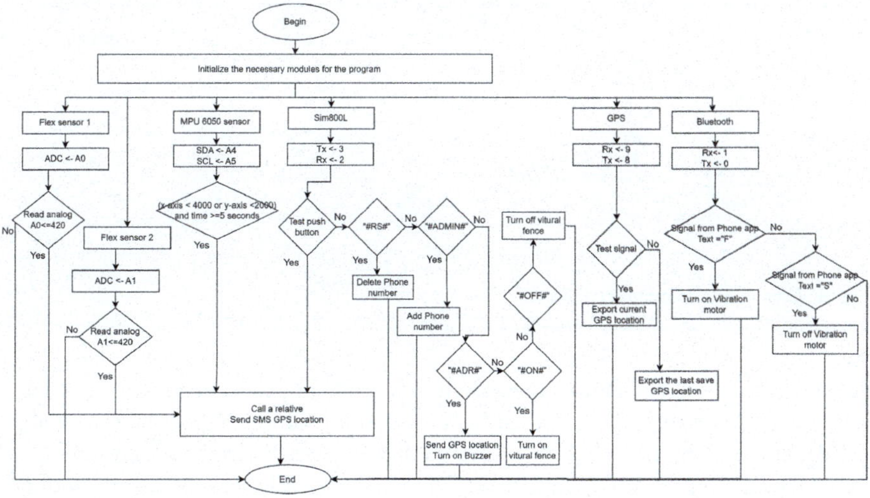

Fig. 3 Segi's algorithm flowchart

will check if the x-axis value is less than 4000 or the y-axis value <2000, and after 5 s, if correct, make a call and send a message to relatives. Next, GPS checks the operation of the Serial pin. If it is correct, it outputs the GPS value at that time. It will send it to the nearest saved location if it is wrong. Finally, the Bluetooth module will check the signal from the phone app if Text = "F" will activate the vibration motor; otherwise, Text = "S" will turn off the vibration motor.

4 Experiments and Evaluation

4.1 Operating Principles

SEGI products are designed like women's bras. So to use the product, we put the shirt on like other normal bra products. Then, we turn on the power switch and reset the phone number of relatives to contact, including the following steps: delete the old phone number previously saved with the syntax #RS# and add a new phone number with syntax #ADMIN#. About 3 s later, the phone number used to set the number of relatives will receive an "OK!" message, which means that the phone number has been changed successfully.

We design using series circuits in 3 lapel positions. The switch will break when a person with bad behavior wants to invade, intentionally takes off a shirt, or struggles. Moreover, suppose a bad guy is an act of squeezing and bending the Flex sensor. In that case, SEGI will send a request for help to the family, including calling and texting GPS coordinates in the form of a link to Google Maps to help relatives locate

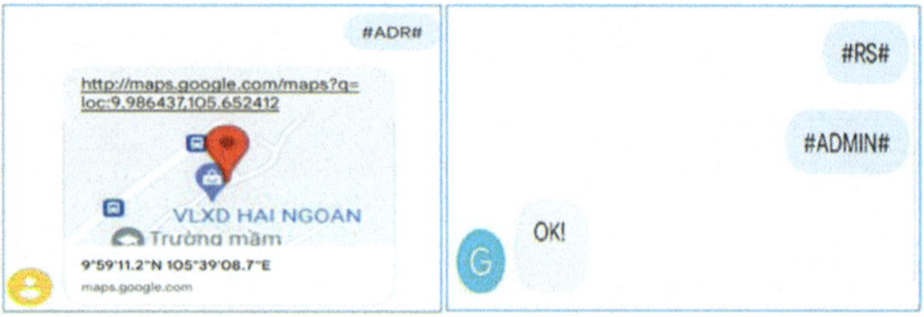

Fig. 4 Check out the message feature

the user's location and shorten the search time when unexpected situations occur. In addition, SEGI is equipped with an SOS emergency button. When the user realizes the danger is approaching, they will immediately press the button. The device will immediately support calling and texting the number of relatives. In case the victim is moved to a new location, relatives can locate the change by texting the phone number installed in the product with the syntax #ADR# as shown in Fig. 4. At the same time, relatives can call the phone number of the sim so that the device makes a sound or help through the buzzer.

The tilt angle sensor device helps to detect when the user has fallen. Moreover, the impact will send a warning to relatives. However, to avoid the false alarm system with the normal fall is not dangerous due to playing jokingly, we installed it after 3 s, the sensor angles are still unchanged, and the device will send an alert to the family.

SEGI can generate a virtual fence. The default radius is 500 m. If they go out of that radius, it will notify the family and helps prevent bad situations in the first place, help family members detect early, and take preventive measures. Start the virtual fence using the #ON# syntax, and turn off the fence using the #OFF# syntax.

The massage feature supports the user's health to avoid breast diseases through the vibration motor. To use this feature, we will install the app on the phone with the interface shown in Fig. 5. Then, turn on the Bluetooth icon to connect with the SEGI device. Next, to start the massage, turn on the first ON icon, and the vibration motor works. Finally, we chose the OFF icon in the middle to finish.

4.2 Evaluation

4.2.1 Design for SEGI

The product is compactly designed with a weight of 220 g and looks like a regular bra for women in Fig. 6. The circuit is designed inside the right coat to ensure the product's aesthetics and not affect the product's quality affect the heart. The capacity of the power source used corresponds to 2000 mAh. Therefore, if the massage feature

Fig. 5 Application interface

Fig. 6 An illustration of product wearing on a model

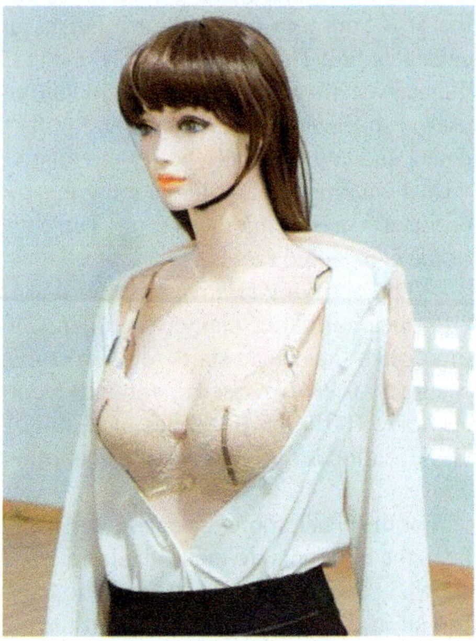

is not used, it can be used continuously for 2 days. On the shirt, there is a circuit to display the battery capacity. Therefore, we can recharge it when ready to run out of capacity.

4.2.2 GPS Positioning and Send a Request for Help to a Relative

GPS navigation circuit GY-NEO 6M V2 is a global positioning module using the US GPS satellite system for fast and accurate positioning speed. GPS works well in all climates and is low-cost. However, several factors can interfere with the signal, cause transmission time loss and reduce accuracy. For example, you are in high-rise buildings with many large trees and stormy weather. The coordinate accuracy of GPS is about 5 m. If the GPS loses signal due to objective conditions, we have saved the last GPS lost position in EEPROM.

4.2.3 Abuse Detection and Notifications for Relatives

To detect abuse, we test the device with the following scenarios. For the first test case, the user feels threatened before that person performs the attacks: They can press the *SOS* button to notify their relatives. In the case the user is suddenly attacked, causing a fall: The device can rely on the tilt angle sensor. If they cannot stand up by themselves after 5 s, the device will automatically notify their relatives. Suppose the user is squeezed into the chest and pulls the bra strap. Again, the Flex sensor will detect and automatically notify the relatives. In addition, we designed a GPS-based virtual walking feature that is set with a radius of 500 m from the location where the user starts messaging the setting. The device will notify relatives if the victim is out of the considered radius.

When notifying relatives, we set it to include calling and texting with the user's GPS location. After many experiments, we have the following results: the slowest time when the device starts sending commands to receive the call is 3.16 s, with an accuracy of 98%. After 1 s, you will receive a message with GPS coordinates, with accuracy 100% for 100 times of testing. However, there are some cases where the GPS location is different from the coordinates at that time, the reason we have shown above.

For the accelerometer sensor, we install it in 5 s with an angle greater than 60 °C from the horizontal to make an alarm. We conduct the test via Serial and get the data shown in Fig. 7. We do not warn if the sensor's height changes, i.e., the user can stand and sit. If they lie, they can turn off the button in the center of the front of the bra.

In some cases, relatives need to change the contact phone number. We deleted the old phone number before and then added a new one with quick and easy texting syntax. On average, in 5 s, we can complete the job of changing the number. In addition, to search for a new device or user's location, after sending a message, receive a GPS location response on average of 3.45 s.

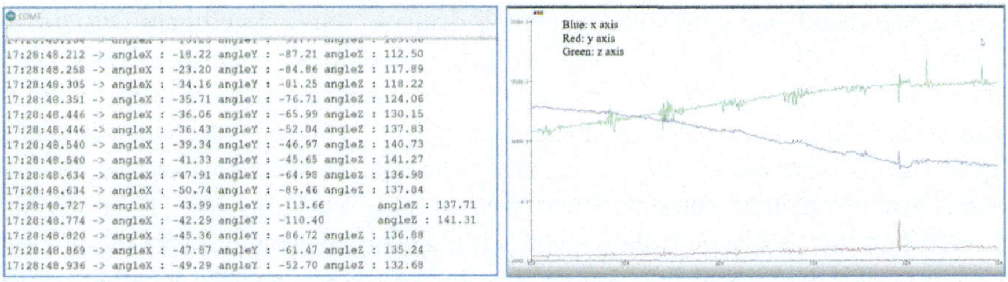

Fig. 7 Check accelerometer sensor via serial

4.2.4 Breast Massage Feature

Massage helps reduce the risk of breast disease. We use a Vibration motor with a speed of 9000 rpm, controlled via an app on a Bluetooth-connected phone. The electricity used is 70 mA, so it only consumes a little power. We invited 10 women to experience this feature. After testing, they give 100% results in comfort and ease of use.

4.3 Comparison with Other Devices

The proposed system should be compared with existing and up-to-date technology. Several parameters need to be considered in order to evaluate the performance of the proposed guidance system. The first parameter is the system's power consumption and uptime without recharging. The criteria are as follows: power consumption of 0–0.5 W is considered low power, 0.5–1 W is medium consumption, and higher than 1 W is high consumption. The second parameter is response time. The system recognizes and responds 0–100 ms is considered fast. An average of 100–200 ms and higher than 200 ms is slow. The third parameter is being able to notify loved ones when in danger and their location. The device is portable if the weight is light, and the user can easily wear it for a long time; otherwise, it is considered non-portable. The fourth parameter finds the new location when lost (Table 1).

Table 1 Comparison between the proposed safety woman device and others

Device	Power	Response time	Alerts	Portable	Find device
Ours	Low	Fast	Yes	Portable	Yes
Smart intelligent system for women and child security [9]	Medium	Fast	Yes	Portable	No
Women safety and protection through smart wearable device using IoT	Medium	Medium	Yes	Portable	No
Smart shoe for women safety	High	Medium	Yes	Portable	No

We compared our performance with 3 devices. Smart Intelligent System for Women and Child Security [9] uses a pressure sensor. As soon as the attacker is about to attack a woman/child, then a regular SMS with the victim's location will be sent to the mobile number of the parent/guardian stored in the device when buying it, then a call. The call will be redirected to the police if the call is not answered. Women Safety And Protection Through Smart Wearable Device Using IoT [10] resembles a smart belt that, when activated, records the sensor values and tracks the victim's location using the Global Positioning System (GPS). The tracked locations of the victims who face harassment can be informed through emergency messages to the victim's contacts. Smart Shoe for Women Safety [11] suggests a smart shoe that helps women take care of themselves but also helps them be fearless. This project uses GPS, GSM modules, and a camera that is interfaced with a Raspberry Pi board and Arduino. With this number of circuits, we estimated the power consumption of smart shoes to be very high.

5 Conclusion

This article focuses on creating simple, cheap, safe products for women. The test has proven the product to be highly accurate against intruders in active and passive situations, with 98% accuracy in low latency calls. At the same time, sending GPS positioning to relatives in the event of a fall or danger or when they are moved to another location can still be found. The product comes with a chest massage feature to comfort the user. The device is easy to use, they do not need to go through any training to use it proficiently, but it only takes a few hours to get used to it at the latest. In the future, the team will improve the removable product with bras so they can be washed again after use.

References

1. D. Sunehra, V.S. Sreshta, V. Shashank, B.U.K. Goud, Raspberry PI based smart wearable device for women safety using GPS and GSM technology, in *2020 IEEE International Conference for Innovation in Technology (INOCON)* (IEEE, 2020), pp. 1–5. https://doi.org/10.1109/INOCON50539.2020.9298449
2. N.R. Sogi, P. Chatterjee, U. Nethra, V. Suma, Smarisa: a raspberry pi based smart ring for women safety using IoT, in *2018 International Conference on Inventive Research in Computing Applications (ICIRCA)* (IEEE, 2018), pp. 451–454. https://doi.org/10.1109/ICIRCA.2018.8597424
3. V. Hyndavi, N.S. Nikhita, S. Rakesh, Smart wearable device for women safety using IoT, in *2020 5th International Conference on Communication and Electronics Systems (ICCES)* (IEEE, 2020), pp. 459–463. https://doi.org/10.1109/ICCES48766.2020.9138047

4. K. Sagadevan, D.S. Kumar, S. Poonguzhali, A. Sivasangari, G. Ilakiya, S. Ragavi, A design of digital tote bag for women's safety, in *2021 10th International Conference on Internet of Everything, Microwave Engineering, Communication and Networks (IEMECON)* (IEEE, 2021), pp. 01–04. https://doi.org/10.1109/IEMECON53809.2021.9689128
5. T. Khandelwal, M. Khandelwal, P.S. Pandey, et al., Women safety device designed using IoT and machine learning, in *2018 IEEE SmartWorld, Ubiquitous Intelligence & Computing, Advanced & Trusted Computing, Scalable Computing & Communications, Cloud & Big Data Computing, Internet of People and Smart City Innovation (SmartWorld/SCALCOM/UIC/ATC/CBDCom/IOP/SCI)* (IEEE, 2018), pp. 1204–1210. https://doi.org/10.1109/SmartWorld.2018.00210
6. A.Z.M.T. Kabir, A.M. Mizan, T. Tasneem, Safety solution for women using smart band and CWS app, in *2020 17th International Conference on Electrical Engineering/Electronics, Computer, Telecommunications and Information Technology (ECTI-CON)* (IEEE, 2020). https://doi.org/10.1109/ecti-con49241.2020.9158134
7. S.R. Mahmud, S.N. Tumpa, A.B. Islam, C.N. Ferdous, N. Paul, T.T. Anannya, BONITAA: a smart approach to support the female rape victims, in *2017 IEEE Region 10 Humanitarian Technology Conference (R10-HTC)* (IEEE, 2017). https://doi.org/10.1109/r10-htc.2017.8289061
8. M.N. Islam, N.T. Promi, J.M. Shaila, M.A. Toma, M.A., Pushpo, F.B. Alam, S.N. Khaledur, T.T. Anannya, M.F. Rabbi, SafeBand, in *Proceedings of the 16th International Conference on Advances in Mobile Computing and Multimedia* (ACM, 2018). https://doi.org/10.1145/3282353.3282363
9. S.K. Punjabi, S. Chaure, U. Ravale, D. Reddy, Smart intelligent system for women and child security, in *2018 IEEE 9th Annual Information Technology, Electronics and Mobile Communication Conference (IEMCON)* (IEEE, 2018). https://doi.org/10.1109/iemcon.2018.8614929
10. S. Valli, S. Kameshwari, Women safety and protection through smart wearable device using IoT. J. Optoelectron. Laser **41**(5), 491–498 (2022)
11. V. Sharma, Y. Tomar, D. Vydeki, Smart shoe for women safety, in *2019 IEEE 10th International Conference on Awareness Science and Technology (iCAST)* (IEEE, 2019). https://doi.org/10.1109/icawst.2019.8923204

Smart Blind Stick Integrated with Ultrasonic Sensors and Communication Technologies for Visually Impaired People

Linh Thuy Thi Pham, Lac Gia Phuong, Quang Tam Le, and Hai Thanh Nguyen

Abstract The inability of blind people to see also causes difficulties when moving on the road, so it is easy to collide with objects. Walking sticks are the most frequently used mobility aids for blind people to move independently and safely. This study introduces a walking stick for the blind with smart, simple, affordable, and friendly criteria. We have deployed it to improve the mobility of both blind and visually impaired people. The main component of this system is an ultrasonic sensor that scans obstacles through reflected waves. The reflected signals from the barrier objects are input to the microcontroller. The microcontroller is then used to determine the direction and distance of objects around the blind. After that, it can notify people how many meters ahead the obstacle is in front by voice, accompanied by vibration. In addition, the system integrates a subscriber identity module card (SIM800L), GPS, and MPU 6050 acceleration sensor. When a blind is in danger, he or she can press the emergency call button. In addition, if he or she falls after 5 s without self-control, the 800L sim can receive a phone call about a pre-installed relative with a message that their GPS location is in danger in the form of a Google map link. At the same time, it warns of danger when the blind move into an area with high gas concentration or large fire around.

L. T. Thi Pham
Can Tho University of Technology, Can Tho, Vietnam

L. G. Phuong · Q. Tam Le
FPT High School, FPT University, Can Tho, Vietnam
e-mail: quangltfsc0378@fpt.edu.vn

H. Thanh Nguyen (✉)
Can Tho University, Can Tho, Vietnam
e-mail: nthai.cit@ctu.edu.VN

N. H. Phuong and V. Kreinovich (eds.), *Deep Learning and Other Soft Computing Techniques*, Studies in Computational Intelligence 1097,
https://doi.org/10.1007/978-3-031-29447-1_11

1 Introduction

According to the World Health Organization (WHO), about 314 million people are blind and have low vision, of which 45 million are blind, and those over 50 forms about 80%.[1] Every 5 s, the world has one more blind person, and every 1 min, a child goes blind; 90% of blind people live in poor and developing countries with difficulty accessing medical services. Meanwhile, according to the report in,[2] there are about 2 million people with poor eyesight in Vietnam, have poor eyesight, a third of which are poor people who cannot afford the treatment that brings light.

A visually impaired person with visual impairment can no longer see the light in both eyes or only dimly distinguish the light shining. Blind people cannot use their eyes to walk but have to fumble around, and the same goes for finding objects in the room. Therefore, they often experience apprehension, fear of dangerous accidents, and unsafety when moving, especially in an environment outside their familiar space. However, the blind also knows how to exploit the signal information about the outside world provided by the remaining senses, taking advantage of all possible conditions. Next is analyzing and synthesizing necessary thoughts and perceptions for their activities. Normally, blind people use all their senses, mainly the three organs of touch, hearing, and smell, to perceive.

In walking, blind people can both walk and perceive their surroundings through these other senses. However, to be safer, they sometimes use bamboo sticks, considered inseparable from the blind, to help them mark their way back and forth. They can detect obstacles and sense the outside environment from the stick. Researching the ability to move footsteps and use canes when walking, the team found that they used the feeling of their feet to sense the slope, bumpiness of the road, slow steps, and no walking. Lift your feet high, swinging the cane sideways to detect obstacles from the knee. These canes often have to be small in size and weight because many visually impaired people are in poor health, move slowly, and have a balanced design. By the way, they usually use their hand, and they can use them for a long time.

Previous studies have focused on using ultrasonic or laser sensors to detect obstacles in front of the blind by transmitting and receiving reflected waves. It generates sound or vibration in response to detected obstacles to warn the blind such as Smart White Cane [1] and Smart Stick [2]. Or integrate the Global Positioning System (GPS) [3] in combination with image processing from the camera.

These solutions still have many disadvantages, such as the inability to detect stairs going down, puddles, and holes. In addition, the feedback to users in the form of vibration and design couldn't be more aesthetically pleasing. Don't alert when they have fallen, have high gas concentration, or have a fire around.

In this article, we propose a smart stick with several contributions:

---

[1] https://www.who.int/news-room/fact-sheets/detail/blindness-and-visual-impairment, accessed on 11 March 2022.

[2] http://t5g.org.vn/viet-nam-co-khoang-2-trieu-nguoi-mu-va-thi-luc-kem, accessed on 11 March 2022.

- We designed the stick to detect obstacles, which can recognize and warn the stairs up and down, left and right.
- We use two means to transmit information to the blind. We have integrated a vibration motor in the stick and a voice alert announcer to notify the user.
- We achieved a very fast response time calculated as 39 ms over an average distance of less than 200 cm before hitting an obstacle.
- When in danger, users can contact their loved ones through the SOS button. Or when they lose consciousness and drop the stick, the system will automatically warn without pressing the button. Instead, relatives will receive an alert with an incoming call from the sim installed on the device and attach the GPS coordinates as a Google map link.
- Relatives can find the cane when lost through an easy text message or directly calling the stick's phone number. The relative's phone number can be changed when necessary.
- Warning of danger when there is a high concentration of gas or fire near the area the user is going to.
- The product is light in weight, does not take much time to train, is aesthetically pleasing, and has enough power to move around for a day.

The rest of this study is organized as follows. Section 2 introduces the main related works. Section 3 presents a brief introduction to the technical requirements, product design, construction of wiring diagrams, algorithmic flowcharts, and operating principles of Safety Stick. Then, Sect. 4 exhibits and analyzes the obtained results. Finally, we conclude the study and discuss future work in 'Conclusion'.

2 Related Work

Several studies related to smart devices to assist the blind have been born to ensure the safety of the blind. For example, this research [4] resulted in a stick with an auditory capability that could detect impediments such as potholes, puddles, and fires while moving around and determine the location point using the Global Positioning System. The study findings revealed that the smart white cane created has an effectiveness of 82% when used with very effective criteria; however, the practicality is around 70% in applicable criteria.

The author [5] has created a system to help the visually-impaired upgrade the white cane by implementing four essential elements. First, an electrical setup employing a simple computer board (Raspberry Pi) connected to an HC-SR04 ultrasonic distance sensor that measures the distance to an object using ultrasonic sound waves and voice feedback that alerts the owner of nearby objects via a speaker. Furthermore, a Raspberry Pi Camera for real-time object detection with a deep learning algorithm, "YOLO," is used. Finally, the locations of these elements in the white cane are in the blind spots to avoid accidents for the owner.

The author [6] makes an integrated tool with the Haar cascade method to detect faces to determine whether there are humans or other obstacles that can show 94% accuracy. In addition, a global positioning system (GPS) can help blind people show the destination's direction. However, it still has drawbacks when crowded conditions cause the camera to continue to take pictures, which will impact the computational weight of Raspberries.

In [7], researchers attached a Radar device with a precise millimeter-level signal to a white walking stick. This device can detect obstacles and distinguish humans from usual obstacles to Anticipate potential hazards. It works by identifying the small displacements of the ribcage caused by physiological activity that is the main sign of a human presence. Smart White Cane [1] is recommended as a Sophisticated and Economic Walking Aid. Their proposed stick detects obstacles to help blind people navigate comfortably. Their device comprises an ATMega328PU microcontroller, four ultrasonic sensor modules HC-SR04, audio IC-APR33a3, a vibration motor, headphones, and batteries with Audio feedback. Their method alerts users with pre-recorded audio messages and haptic feedback in the form of vibrations. The cane can detect holes, potholes, falls, stairs (up and down), low and knee obstacles, and even obstacles higher than the waist. The battery compartment is hidden in the bar, thus reducing the risk of damaging the circuit and making the device less bulky. In addition, the system offers an ON/OFF switch, vibration feedback, and an audio jack on the handlebar. However, the system has no locating method to find the user's location, cannot detect oncoming vehicles and slippery floors, and has no fire or smoke alarms.

The author [8] designed and developed a smart cane capable of sensing and signaling the environmental features around visually impaired people. The proposed smart cane has two ultrasonic sensors for measuring the distance between the smart cane to the stationary and hanging objects found to be an obstacle. The actuator provides an alert signal when the obstacle falls inside a threshold level of 2 m. The design integrates a camera module ESP-32 to capture moving images of objects/people and obstacle detection. The camera module gets an enabled signal from the PIR sensor to capture the image of the moving object/person and save processor time.

In general, we find that these products and devices are often expensive, and it is not suitable for the psychophysiology and habits of the blind when using them. For example, user guides and indicators often use sound and vibration signals on the back, shoulders, and handles. In addition, bulky, energy-intensive, noisy equipment affects people and the user's ability to perceive the environment. In addition, the studies focused on obstacle avoidance and GPS positioning but did not study warnings to loved ones when in danger.

3 Methods

3.1 Requirements

The device does not interfere with normal sensory habits for walking, such as requiring the user to wear gloves or interfering with environmental sounds.

The device has a small weight. The total weight should be at most 1 kg, which helps users to use it for a long time without hand fatigue. Each charge for about 5 h allows about 1 week of use before having to charge the battery if, on average, the blind person walks for about 30 min daily. Or enough battery capacity so they can last all day outdoors.

About obstacle detection accuracy: must be greater than 90%-obstacle detection time: less than 0.3 s. The sensor must detect obstacles from the tip of the foot upwards to the level of the user's abdomen, at a distance of 0.5–2 m, so that the user has time to react and not be confused with objects far.

When the visually impaired are in danger, they can call their relative via pre-set phone numbers, including their location coordinates on SMS messages. When the cane is lost, relatives can assist in finding the stick through the phone sim on the stick.

Flame sensors need to detect fire when there is a real fire near the user or when in an area with high gas concentration and warn them of danger. The product must be flexible, the components of which can be replaced or upgraded as needed. The price is as low as possible to suit the economic conditions of most blind people in Vietnam.

The device can be used after 10 min of training. Tool proficiency levels can vary, but users can generally feel confident in less than 2 weeks of use. All blind people can use it easily through training time, suitable for adults and people who are blind due to advanced age.

3.2 Design for the Stick

The blind cane has a shaft length of 124 cm, a folding length of 33 cm, an outer diameter of the tube is 13 mm, and a wall thickness of 1mm. Stick material is made of aluminum alloy, non-oxidation, electrical insulation, waterproof, non-fading, easy to clean, non-slip, insulated, lightweight, weight 100 g (without equipment), and 300 g with the power box. Colors include red and white. The handle elastic rope is made of rubber and has the characteristics of resilience, strong tensile strength, and aging resistance. The upper end of the handle is fitted with a flexible and easy-to-replace wrist strap. The center console box is designed in a box shape with dimensions of $115 \times 90 \times 55$, white color. ABS plastic material to help prevent water and good protection for the power supply. The design includes 2 control boxes above and below. The top includes 1 Lipo battery 3.7 V–1000 mAh, GPS, sim 800L, MPU

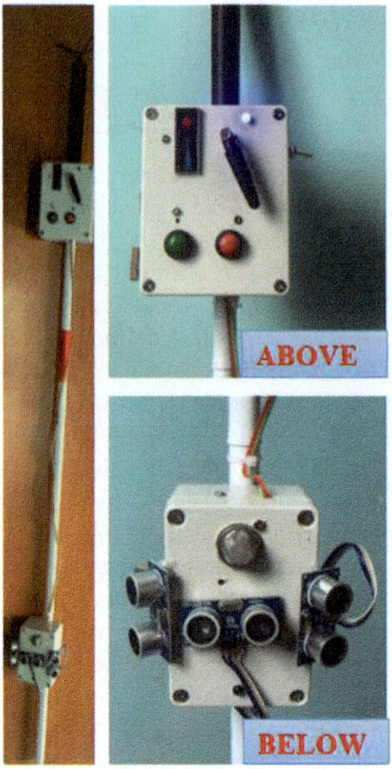

Fig. 1 Design of safety stick

6050 sensor, battery capacity display circuit, SOS button, reset button for relatives phone number, power on/off button located on the side on the box 1 board Arduino Nano. The bottom includes 3 ultrasonic sensors placed 3 front, left, and right corners, 1 gas sensor, 1 flame sensor, 1 DFPlayer module with the memory card, speaker, 1 Vibration motor, and 1 Lipo battery 3.7 V–1000 mAh, and 1 board Arduino Nano (Fig. 1).

The circuit diagram of the product is shown as Fig. 2 including Arduino Nano acts as a central processing unit, receiving signals from SIM 800L, processing signals, and outputting signals (Digital, PWM) output to implement devices (vibration motor, buzzer, speaker). The power source includes 2 Lipo batteries with 3.7 V–1000 mAh. Arduino mini vibration motor when input is High, the motor vibrates. Three ultrasonic sensors, HY-SRF04, emit an ultrasonic wave and then measure the time from emission to return of the wave to measure the distance from the position current to the object. SIM module 800L reads the status of the emergency button. If the button is turned on, the sim will send messages and calls to the pre-set relatives. GPS will locate the coordinates of the blind. The switch is used to turn the device on and off. The acceleration sensor is based on the tilt relative to the axis to warn of a fall when necessary. Voice and speaker circuits are used when receiving a signal from the ultrasonic sensor, acceleration sensor, or SOS button. It will make a sound to warn blind people.

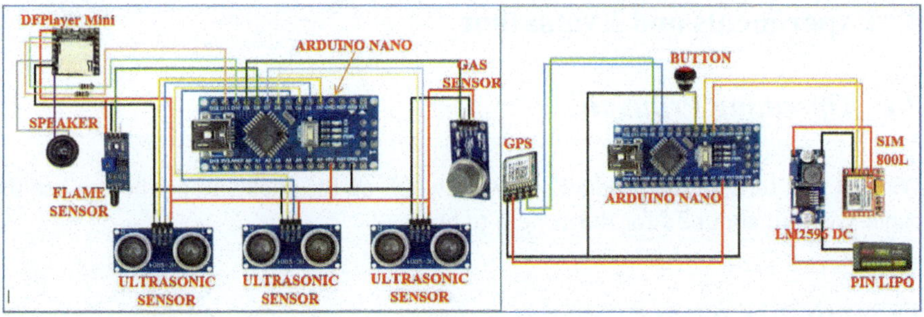

Fig. 2 Circuit diagram of safety stick

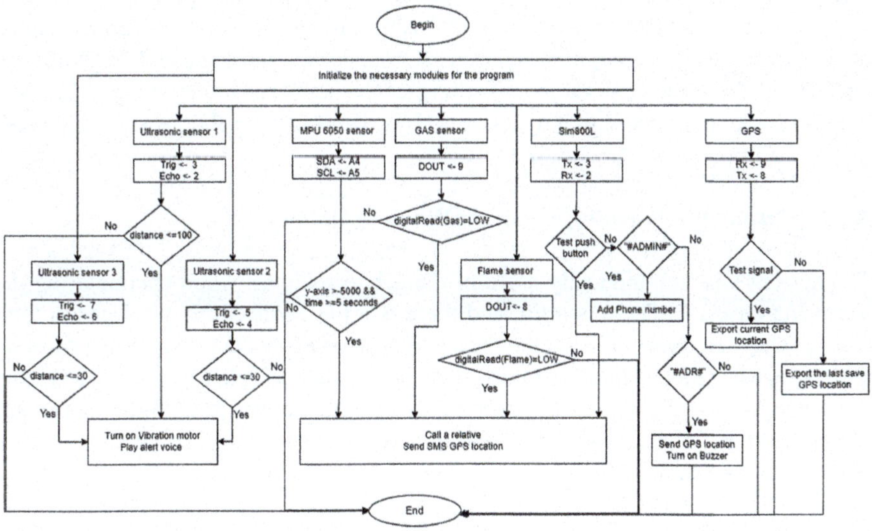

Fig. 3 The workflow for safety stick

The flowchart is shown in Fig. 3. When the switch is turned on, the Arduino Module will be initialized. Inputs are ultrasonic sensors that will continuously check the signal for obstructions. If there is an obstacle between 20 and 200 cm, it will activate the vibration motor and speaker. Distances greater than 200 cm count as out-of-coverage areas. The fire sensor input will be checked, if there is a fire, the motor will vibrate, and the speaker will work. The SIM 800L module will check the activation status of the push button. If activated, the device will call and send GPS location messages to the pre-set phone number. In addition, when the stick is lost, relatives can send a text message to the phone number in sim800L.

4 Experiments and Evaluation

4.1 Operating Principles

We will first turn on the product's power switch on the stick's upper box to use the cane. We wait about 2 s for the system to boot.

4.1.1 Obstacle Avoidance Function

The product includes 3 front ultrasonic sensors. When the blind person moves from 20 to 200 cm away from the obstacle, the vibrating motor will work, accompanied by a voice broadcast through the speaker, "Obstacle is ... meters away", signaling an obstacle. If the obstacle is far from them (greater than 2 m), the motor vibrates, and the speaker stops because that distance is considered safe for the visually impaired.

4.1.2 Automatic Warning Function When in Danger

The system will automatically warn of danger to relatives when encountering the following cases: 1. There is a fire and there is a high concentration of gas in their area. 2. They fell, but within 5 s, they still couldn't get up on their own. When warning of danger, we install a vibrating motor and a buzzer that will alert the user in advance. After that, the system will call back to relatives with a message of GPS coordinates that they are in danger in shown Fig. 4. In addition, users can press the SOS button when they need support from their relatives.

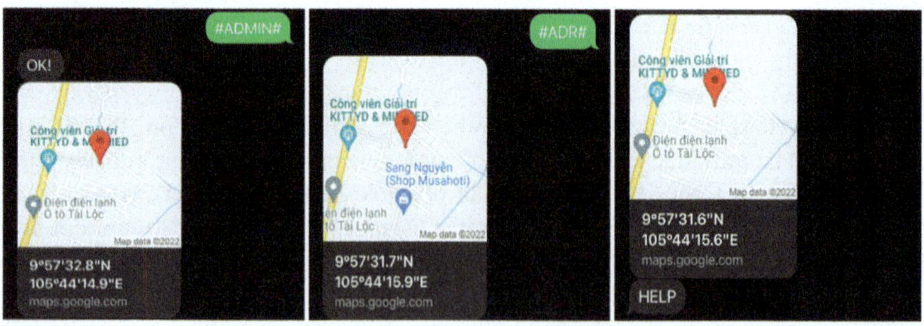

Fig. 4 Sending a message including GPS coordinates in emergency cases

4.1.3 Finding a Lost Stick

When the cane is lost, relatives can send a text message to the phone number installed in SIM800L with the syntax: #ADR#, and the relative will receive the location coordinates where the cane is lost. If relatives have lost or changed their phone number, they can press reset and text with the syntax #ADMIN# to change the phone number.

4.2 *Evaluation*

4.2.1 An Overview on Design

The device does not interfere with the usage habits of the visually impaired, and lightweight helps them to use it for a long time without hand fatigue. Furthermore, aesthetic design, easy to use, does not take too much time to train and is suitable for all blind people, such as the elderly and children.

4.2.2 Detecting Obstacles and Navigation

With a design including 1 sensor HY-SRF04 designed in front of the cane with Servo to help blind people receive warnings about dangers from many directions. The advantages of ultrasonic sensors are fast processing and relatively accurate results. So we tested an obstacle between 20 and 200 cm from the clubfoot with 100% accuracy on all tests. However, its disadvantage is that due to the use of ultrasonic waves and their reflection to calculate the distance and detect obstacles, it generates signal noise that makes the club unable to stop vibrating or the whistle to howl continuously (probability 1% after tests). When testing with different types of obstacles, such as bricks, wood, soft towels, glasses, and transparent objects (glasses), in 10 different tests (with a distance of 20–100 cm from the club foot). The result is shown in Fig. 5.

4.2.3 The GPS Positioning Feature

We use the NEO-6N Module to locate the location coordinates of the blind. Results are sent quickly and accurately. However, this device will be affected when blind people are in high-rise buildings, large trees, or stormy weather. To overcome this drawback, we have saved their last position before the signal loss into EEPROM.

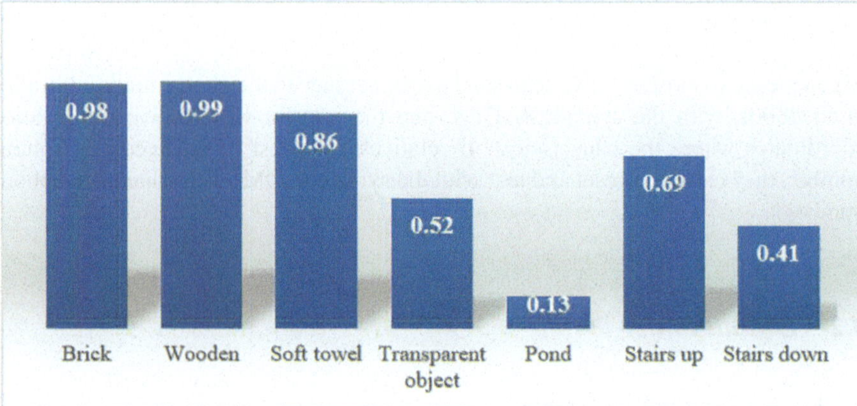

Fig. 5 Obstacles detected by ultrasonic sensors

4.2.4 Sending Warnings and Notification in Danger Cases to Relatives

When notifying a loved one, we set it to include calling and texting with the user's GPS location. After many experiments, we have the following results: the slowest time when the device sends commands to receive the call is 3.25 s, with an accuracy of 98%. After 1 s, you will receive a message with GPS coordinates, with an accuracy of 100% for 50 times of testing. However, there are some cases where the GPS location is different from the coordinates at that time, the reason we have shown above. Figure 6 is recorded by us in one of the dangerous warning SOS button tests.

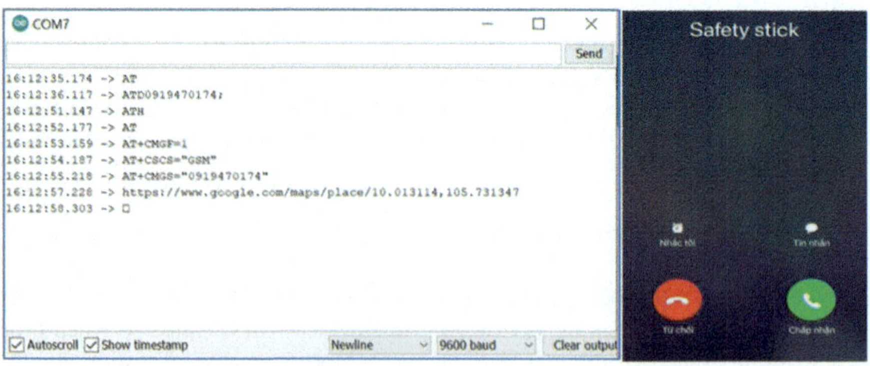

Fig. 6 Sending a message of SOS via serial

4.2.5 Find Lost Stick

When a relative sends a text message with a specified syntax to the pre-set phone number, they can receive a message revealing the location of the lost stick along with a Google link map. The average time after 50 test cases is 3.35 s, the accuracy is 100%.

4.2.6 Regarding Danger Warning in Areas with Fire and High Temperature

The product has a fire sensor installed using an infrared light head (the infrared head is exposed from the electrical box). Normally, the sensor will identify infrared rays in the surrounding environment, and if there are dark red rays, a wavelength in the range of 760–1100 nm from the light source will alert the user. However, this sensor can be easily damaged when exposed to excessive heat. To be sure, our team tested with a small spark by a lighter with a distance ranging from 1 to 100 cm. The results obtained have an average accuracy of 95% and an average response time of 22 μs.

4.2.7 Power Capacity

Stick uses 2 Cell Lipo Battery 3.7 V–1000 mAh. Ultrasonic sensor current consumption is 40mA, Arduino Nano board 40 mA, Sim800L 10 mA standby, GPS Neo 6 M 67 mA, MPU 6050 accelerometer 3.6 mA, and some other materials ranging from 5 to 40 mA. Therefore, the cane can be operated continuously for more than 1 day. Enough time if blind people to do outdoor activities. Reusable rechargeable battery to save users.

To evaluate the performance of this stick, we tested it on actual beneficiaries (trained and untrained) blind people. Six blind people at The School experimented on disabled children in Can Tho, Vietnam. Two were trained to use a cane, and four were not.

During the experiment, a blind person was asked to walk through the testing area where 5 different types of obstacles, including rocks, trees, steps, glass doors, and puddles, were placed within a 15 m range. First, the user's walking speed is recorded. Next, the time taken by trained and untrained to successfully walk through the obstacles is measured as shown in Fig. 7. Then we calculated the average speed of trained users to be 0.77 m/s and untrained users to be 0.45 m/s. In comparison, the traveling speed of the sighted people is (1.4 m/s). This result shows that training the user reflected in gaining twice traveling speed and increased the user's trust in avoiding obstacles in the free path.

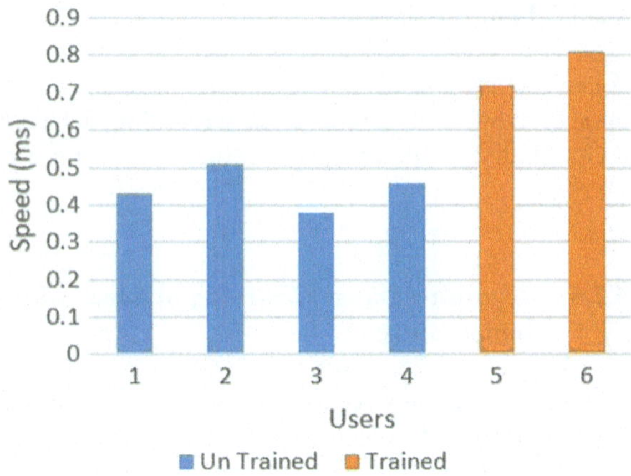

Fig. 7 Performance and accuracy of the Safety stick

4.3 Comparison with Other Devices

The proposed system should be compared with available and up-to-date technology. Several parameters should be considered to evaluate the performance of the proposed guidance system. The first parameter is the range of detection. A tool that can find obstacles throughout 0–2 m can be considered a low-range device, 2–4 m as medium range, while higher than 4 m is considered a high range. The second parameter is the response time, and a system sensing and responding 0–100 ms is regarded as fast, 100–200 ms medium, and higher than 200 ms as slow. The third important parameter is the system's power consumption and how long it will stay working without recharge. The following ratings are considered: electrical power consumption of 0 0.5 W is regarded as low power, 0.5–1 W as medium consumption, and higher than 1W as high consumption. Fourth, the device is portable if it is light in weight, and the user can easily wear it for an extended period otherwise. It is considered non-portable. The fifth parameter finds the stick when it is lost. Users often hold the cane in a standing or inclined position. If the cane falls, they are also counted as having a fall requiring a warning. So the sixth parameter we compare is the fall warning.

We compared our performance with 3 devices. The first one proposed a Low-Cost Smart Cane using GPS [9] consisting of two sensors: a water sensor, an ultrasonic sensor, and a microcontroller (Arduino Uno R3) to receive the sensor signals. RF Module helped to find the stick if it was lost in a certain range. The GPS navigation system in the Mobile could be used to guide blind people to find any places. We also compare to a blind guidance system [10], including an ultrasonic sensor employed to detect obstacles and potholes within 2m. Moreover, GPS and ESP8266 Wi-Fi Module were utilized to share location with the cloud. The MQ2 Gas Sensor is used to detect the flame in the path, and an RF Tx/Rx module is used to find the smart cane when it is misplaced. The third one for the comparison included a Vibration and Voice Operated Navigation System for Visually Impaired Person [11] using an

Table 1 Comparison between the proposed safety stick and others

Products	Detection range	Time response	Alerts	Portable	Find stick	Fall warning
Ours	Mid-range	Fast	Yes	Portable	Yes	Yes
Development of low cost smart cane with GPS [9]	Mid-range	Fast	No	Portable	Yes	No
BlinDar [10]	Mid-range	Fast	No	Portable	Yes	No
Vibration and voice operated navigation system [11]	Low	Low	No	Portable	No	No

ultrasonic device to determine the range of obstacles and also a microcontroller to act accordingly. The system includes a warning system through voice rendering and through the generation of vibration. The comparison is shown in Table 1.

5 Conclusion

This paper focuses on the design of a simple, low-cost smart cane for navigation for the blind. Through the tests, the product has proven to be highly accurate in detecting obstacles, fire, and dangerous gas near the user (the most accurate is less than 2 m). Besides, the product achieves 98% accuracy in calls made via the SOS button with low latency. The average time from pressing a button to making a call was 3.25 s for 50 tests. Simultaneously send GPS positioning to loved ones in the event of falls, danger, or lost sticks with high accuracy. Because of the product's simplicity, relatives or visually impaired people only need to undergo training to use it proficiently, but only a few hours at the latest. In the future, we will develop to avoid obstacles with many materials for high accuracy, such as glass, wood, slippery surfaces, and warning stairs to go down. Furthermore, integrate more Camera Module to be able to know the problems encountered by the blind for relatives at home to monitor. The product was tested by students at The School for disabled children Can Tho, Vietnam.

References

1. R. Sheth, S. Rajandekar, S. Laddha, R. Chaudhari, Smart white cane-an elegant and economic walking aid. Am. J. Eng. Res. **3**(10), 84–89 (2014)
2. M.P. Agrawal, A.R. Gupta, Smart stick for the blind and visually impaired people, in *2018 Second International Conference on Inventive Communication and Computational Technologies (ICICCT)* (2018), pp. 542–545

3. M.P. Agrawal, A.R. Gupta, Smart stick for the blind and visually impaired people, in *2018 Second International Conference on Inventive Communication and Computational Technologies (ICICCT)* (IEEE, 2018), pp. 542–545
4. Y. Imam, M. Hanafi, M. Yasin, S. Yuyus, W. Sapto, H. Zainudin, et al., Web based smart white canes for blind or visually impaired people on wetland areas (2021)
5. A. Carranza, A. Baez, J. Hernandez, H. Carranza, H. Rahemi, Raspberry pi and white cane integration for assisting the visually impaired, in *International Conference of Control, Dynamic Systems, and Robotics* (Avestia Publishing, 2022). https://doi.org/10.11159/cdsr22.126
6. A. Abdurrasyid, I. Indrianto, M.N.I. Susanti, Face detection and global positioning system on a walking aid for blind people, in *Bulletin of Electrical Engineering and Informatics* (Avestia Publishing, 2022). https://doi.org/10.11591/eei.v11i3.3429
7. E. Cardillo, C. Li, A. Caddemi, Millimeter-wave radar cane: a blind people aid with moving human recognition capabilities. IEEE J. Electromagn., RF Microw. Med. Biol. 1–8 (2021). https://doi.org/10.1109/jerm.2021.3117129
8. T.L. Narayani., M. Sivapalanirajan., B. Keerthika., M. Ananthi., M. Arunarani, Design of smart cane with integrated camera module for visually impaired people, in *2021 International Conference on Artificial Intelligence and Smart Systems (ICAIS)* (IEEE, 2021). https://doi.org/10.1109/icais50930.2021.9395840
9. F. Ahmed, Z. Tasnim, M. Rana, M.M. Khan, Development of low cost smart cane with GPS, in *2022 IEEE World AI IoT Congress (AIIoT)* (2022), pp. 715–724
10. Z. Saquib, V. Murari, S.N. Bhargav, BlinDar: an invisible eye for the blind people making life easy for the blind with internet of things (IoT), in *2017 2nd IEEE International Conference on Recent Trends in Electronics, Information & Communication Technology (RTEICT)* (IEEE, 2017). https://doi.org/10.1109/rteict.2017.8256560
11. N. Mahmud, R. Saha, R. Zafar, M. Bhuian, S. Sarwar, Vibration and voice operated navigation system for visually impaired person, in *2014 International Conference on Informatics, Electronics & Vision (ICIEV)* (IEEE, 2014). https://doi.org/10.1109/iciev.2014.6850740

Segmentation of the Abnormal Regions in Breast Cancer X-Ray Images Using U-Net

Nguyen Hoang Phuong, Ha Manh Toan, Nguyen Van Thi, Ngo Le Lam, Nguyen Khac-Dung, and Dao Van Tu

Abstract We present a segmentation of the abnormal regions in breast cancer X-Ray images using U-Net.The breast cancer X-Ray images dataset was collected by the Vietnam National Cancer Hospital radiologists and consists of three image classes: Normal, Benign, and Malignant. The hospital radiologists labeled the dataset's images and performed the localization of lesions of the Benign and Malignant images. We use the breast cancer X-Ray image dataset to train the U-Net model for segmenting Benign and Malignant masses to visualize Benign and Malignant areas in the images. The images of Normal, Benign, and Malignant classes are converted into a PNG 16-bit format. Before model training, the data pre-processing is realized. We flip the images of R-CC and R-MLO views horizontally so that all breast images are leftward-oriented. Then, we crop the image by probing along the edge of the breast. After that, this image is resized into 512×512 pixels. Our network achieves an average Dice of 0.74357. We have compared segmentation masks of model prediction with doctors' annotation masks. The result shows that the segmentation model has performed well. Finally, our system combines the segmentation result with the heatmap result to have more supporting information on abnormal region localization.

Keywords U-Net · Abnormal region · Breast cancer · X-Ray images

N. H. Phuong (✉)
Artificial Intelligence Division, Information Technology Faculty, Thang Long University, Nghiem Xuan Yem Road, Hoang Mai District, Hanoi, Vietnam
e-mail: nhphuong2008@gmail.com

H. M. Toan
Institute of Information Technology, Vietnam Academy of Science and Technology, Hanoi, Vietnam
e-mail: hmtoan@ioit.ac.vn

N. Van Thi · N. Le Lam · N. Khac-Dung · D. Van Tu
Vietnam National Cancer Hospital, 30 Cau Buou, Tan Trieu, Thanh Tri, Hanoi, Vietnam
e-mail: khacdung.nguyen@bvk.org.vn

D. Van Tu
e-mail: daovantu@hmu.edu.vn

N. H. Phuong and V. Kreinovich (eds.), *Deep Learning and Other Soft Computing Techniques*, Studies in Computational Intelligence 1097,
https://doi.org/10.1007/978-3-031-29447-1_12

1 Introduction

In breast cancer screening, localization of the abnormal masses helps breast cancer radiologists who read the X-Ray localize the position of the Benign and Malignant masses in patient diagnosis and therapy. Different localization methods, such as segmentation, heatmap, and border detection, are used to do this task. Each method has its advantages and disadvantages. Therefore the segmentation method provides the most information for radiologists. This information consists of the relative positions of the malignant and benign regions in the image space and details the border representing the borderline between the tumor and the normal tissue regions. But segmentation methods usually perform effectively when objective regions are contrastive versus around. So some methods can achieve low scores in some contexts of small and low-contrastive objective regions. The breast cancer X-Ray image is in this case. To support abnormal region localization information, we combine the segmentation result with the heatmap result in our system. Heatmap is often a tool to help users observe the locations where objects of interest are likely to appear, helping them have an initial view of the places to look. In the context of a diagnostic aid, a heatmap is also a good fit to provide an additional perspective for the physician, even if it doesn't show as much discrimination as a segmentation results. To visualize the abnormal regions as benign and malignant in breast cancer X-Ray images of the mammograms, we present an approach of a segmentation of the abnormal regions in breast cancer X-Ray images using U-Net. We use the breast cancer X-Ray image dataset provided by the Vietnam national cancer hospital radiologists with Benign and Malignant localization to train the model.

To summarize, the main contributions of this paper are:

- We develop an image dataset of Normal, Benign, and Malignant categories.
- We train a U-Net model on this image dataset for segmenting the benign and malignant tumors to visualize a hurt area in breast cancer X-Ray images.
- We visualize the abnormal regions by combining the segmentation and heatmap results.

The rest of the paper is organized as follows: Sect. 2 reviews some related works. Section 3 presents a segmentation model of the Abnormal Regions in Breast Cancer X-Ray images. Section 4 describes a model evaluation, and Sect. 5 gives some conclusions.

2 Related Works

Some first works about deep learning networks for segmenting semantic images using full connected networks (FCN) are described in [13]. FCNs consist of many convolutional layers and have input as one image and output with a segmentation result of the same size as the input. This work played the role of the first work to train CNN end-to-end for pixel-wise prediction. This work also established the basis for many other types of research using deep learning networks for image segmentation by end-to-end approach. One such work is ParseNet of Liu [14], which was improved by adding global semantic information into FCNs. In some works used in medicine, FCNs are applied to some medical problems, such as brain tumor segmentation [15], skin lesion segmentation [16], and iris segmentation [17]. These works focused on applying FCN in particular contexts, such as using the cascade framework of FCNs or requirements of no pre-processing and post-processing. Some models for the segmentation of medical images are developed based on FCNs and the encoder/decoder architectures. For example, U-Net: Convolutional networks for biomedical image segmentation [18] and V-Net: Convolutional networks for biomedical image segmentation [19]. U-Net is designed by modifying and extending FCN to work with few training images and yield more precise segmentations. Some other works were developed and used U-Net architecture to segment different image modalities. For example, in [20], Z. Zhou et al. used a nested u-net architecture for medical image segmentation, and in [21], O. Cicek applied 3D U-net for learning dense volumetric segmentation from sparse annotation. Concerning breast cancer problems, Vargas et al. [10] proposed an approach to Space-occupying lesions identification in mammary glands. In this study, an original image is processed by the arithmetic statistic filter and segmentation algorithms based on threshold techniques and region increaser. Another study by Ittannavar et al. [11] about the segmentation of breast masses in mammogram images using a multilevel multiobjective electromagnetism-like optimization algorithm includes the advantages like enhanced robustness to hold the image details and adapt to the local context. In this study, the authors used two mammogram datasets DDSM and MIAS.

The other way to present the appearance of objective regions is heatmap-based localization. This method is a technique that describes the location and size of regions of interest in a digital image using similarly sized images with the input image, where this image is colored pixel by pixel to represent the probability of a lesion occurring at that pixel respectively. That color representation is equivalent to a probabilistic image, where the value at each pixel implies a warning about the likelihood of the object of interest occurring. Phuong et al. [1] proposed a method to generate a heatmap by patch classification using a deep learning network in the context of breast cancer X-Ray images.

3 A Model for Segmentation of the Abnormal Regions in Breast Cancer X-Ray Images

3.1 Data Collection and Labeling

Collecting and labeling data is the top priority in training deep neural network (DNN) models as it enables DNN models to accurately classify abnormal regions in breast

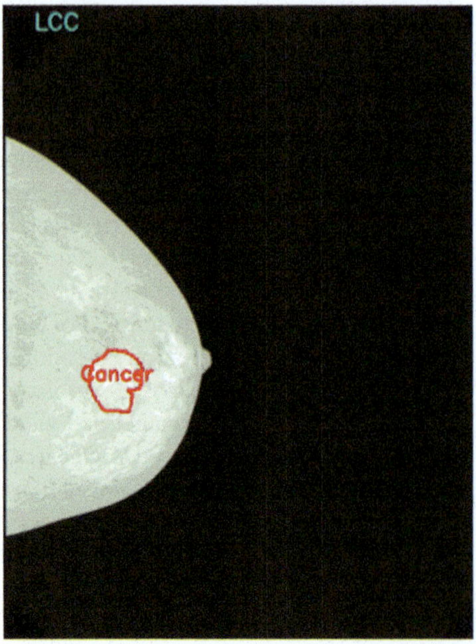

Fig. 1 Two radiologists using the PMR tool localize the abnormal border

cancer X-Ray images. The dataset of breast cancer X-Ray images used in this study was collected by the Vietnam National Cancer Hospital radiologists. It consists of three image classes: Normal, Benign, and Malignant. Each image in the dataset belongs to one of four images corresponding to the four standard views used in screening mammography: R-CC (right craniocaudal), L-CC (left craniocaudal), R-MLO (right mediolateral oblique), L-MLO (left mediolateral oblique). The hospital's radiologists performed the localization of lesions of the Benign and Malignant images by using the PRM software of the hospital. Currently, the total number of the images of Normal, Benign, and Malignant images collected by radiologists is 3431 images, consisting of 1898 Normal images, 821 Benign images, and 712 Malignant images. Figure 1 shows an example of localization of the cancer border.

3.2 Pre-processing and Model Training

First, raw collected data need to be cleaned during the data cleaning process: unqualified, duplicated, and irrelevant images are excluded from DNN models. Data cleaning is crucial to improving data quality and requires an extended processing time.

Input data are DICOM breast cancer X-Ray images. These images of Normal, Benign, and Malignant classes are converted into a PNG 16-bit format. Subsequently, these images are realized by the pre-processing process. Data pre-processing process

Fig. 2 Main steps of pre-processing breast cancer X-Ray images

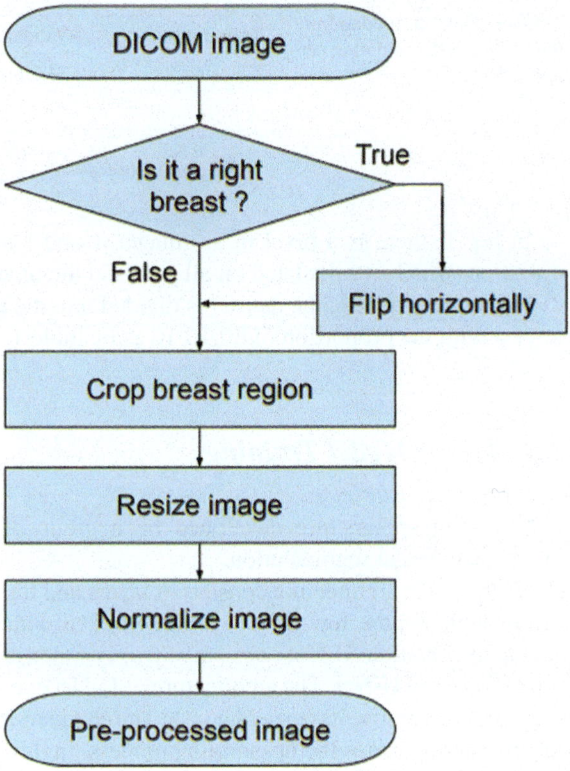

is described in Fig. 2. Firstly, we flip the images of R-CC and R-MLO views horizontally so that the breast is always on the same side of the image. Secondly, the breast region is divided by image processing technique: we crop the image by probing along the edge of the breast. After that, this image is resized into 512 × 512 pixels.

Step 1: We flip the images of R-CC and R-MLO views horizontally so that all breast images are leftward-oriented.

Step 2: We checked the hospital's DICOM images received from radiologists; if there were any letters such as L-CC, R-CC, L-MLO, R-MLO, or any other "noise" then we "clean" the above letters.

Step 3: We cropped images by probing along the edge of the breast to keep the most breast region used in model training.

Step 4: After that, this image is resized into 512 × 512 pixels to fit the deep neural network models when the number of images used for model training is enormous.

Step 5: Finally, each image is normalized.

The formula for normalizing images is presented in Eq. 1.

Table 1 Dataset divided into the training set, validation set, and test set

Training set	Validation set	Test set
2741images	343 images	347 images

$$X_{i,j}norm = \frac{x_{i,j} - \mu}{\delta} \tag{1}$$

In Eq. 1, there is a pixel in an image, μ and δ are the mean and the standard deviation, which is calculated on all pixels in the image.

For model training, the dataset is divided into the training set, validation set, and test set with the proportion of 80:10:10 as in Table 1.

3.3 U-Net Model Training

U-Net is an architecture developed by a group of Ronneberger et al. [18] for biomedical image segmentation.

In Fig. 3, the architecture consists of layers and forms an encoder-decoder design. In this architecture, low-level layers are linked with high-level layers for utilizing spatial information. U-Net consists of two symmetric branches of the letter U therefore, it is called U-Net. The architecture of U-Net consists of two parts: left and right. Each part will realize its particular task. The encoder side compresses the features, and the decoder performs the upsampling process. In this study, the structure of DynamicUnet uses the library Fastai in [2]. The encoder is implemented using the architecture of ResNet 18 and is pre-trained by ImageNet. The decoder part is generated corresponding to the encoder with upsampling using the PixelShuffle technique [3].

The input of U-Net is an image, and the output is a mask image where each pixel is labeled by concrete class. In this case, the input is an image of 512 × 512 pixels, and the output is an image of 512 × 512 pixels, that each pixel of the output image takes one among three values in {0,1,2}. The value 1 is labeled Benign, the value 2 is labeled Malignant, and the value 0 is labeled other cases.

For the training process, the mask images are generated from the segmentation results of radiologists for the input data.

Figure 4 shows a variation graph of the loss values of the training set and validation set in the training process. From the graph, we observe that the training process achieves an optimization position in epoch 30; after that, the validation loss shows no sign of decreasing. Therefore, the model at epoch 30 is optimal, and this model uses for an experiment on the test set. Training time for each epoch is about 15 min.

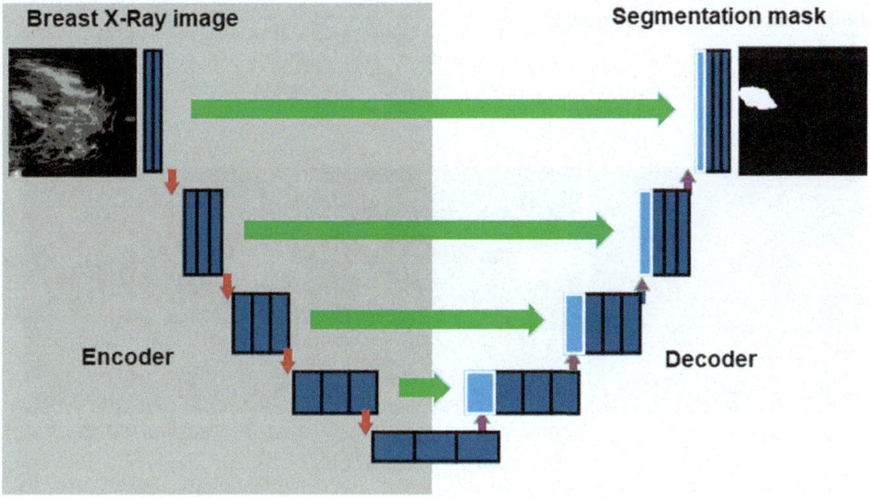

Fig. 3 Architecture of the U-Net model based on experiments in [12]

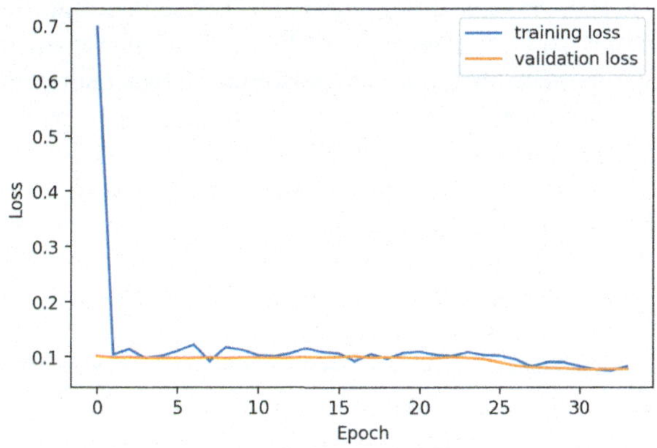

Fig. 4 Variation graph of the loss values of the training set and validation set in the training process

4 Model Evaluation

The trained model is evaluated on the test set to measure the model quality. The output of this step is a Dice metric to access the quality of the model. We have conducted experiments to test the model on the pre-defined test set for the segmentation task. For the breast models [4], our model achieves a 0.743457 average Dice score on the test set. See more details in Table 2. The result shows that the segmentation model has performed well. For the cancer area segmentation task, the model quality is evaluated by comparing segmentation masks of model prediction with doctors' annotation masks on the pixel level. A Dice score metric is selected to evaluate segmentation.

Table 2 Evaluation results on the test set

Metrics	Benign	Malignant	Other	Average
Dice	0.647619	0.598147	0.984606	0.743457

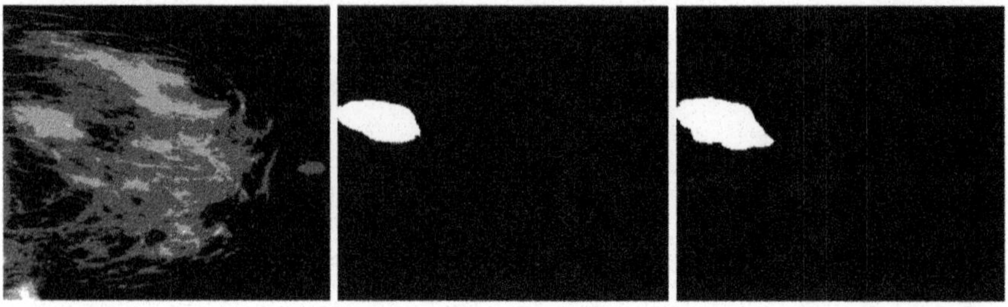

Fig. 5 Comparing segmentation masks of model prediction with doctors' annotation mask

There is some difference in the Dice scores of the three labels, and its reason might come from the imbalance of benign and malignant areas in the dataset. Moreover, the number of normal areas is much larger than benign and malignant. We would review some studies to discuss more Dice in the medical context. In [10], by using a semi-automatic approach, the authors generated the maximum Dice score was 0.92 for the segmentation task of breast cancer MRI images. Here, it is necessary to point out that Dice is a metric with values between zero and 1, and this metric is better when its value is closest to 1 [12]. In a medical image segmentation context, this means that the manual segmentation and the automatic one match when the Dice is 1 and do not match when the Dice is zero. In this sense, typically, values of Dice over 0.75 are good in the medical routine. Figure 5 shows an example of the result of malignant segmentation.

In Fig. 5, from left to right, (a) is an input image, (b) is a segmentation mask of model prediction, and (c) is a doctors' annotation mask. There are some approaches to visualize the lesion regions, such as segmentation, heatmap, or using a rectangle. In most cases, segmentation is one of the most meaningful options. As seen in Fig. 5, we can easy to compare the results of the prediction versus the doctor, even if we need the detail on the border. The Dice score value is 0.853164, a measure used when comparing segmentation masks of model prediction with doctors' annotation masks. Dice is calculated by the formula $Dice = \frac{2|X \cap Y|}{|X|+|Y|}$, where X is the white image points in the segmentation mask and Y are the white image points in the doctors' annotation mask.

Lastly, to support abnormal region localization information, we combine the segmentation result with the heatmap-based abnormal area localization result [1] in our system.

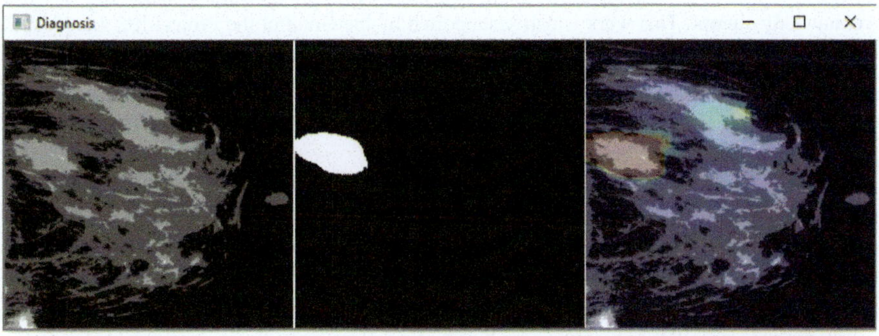

Input image (a) Segmentation mask (b) Abnormal area localization (c)

Fig. 6 An example of an abnormal region in breast cancer X-Ray combining the segmentation and heatmap results

In Fig. 6, from left to right: (a) is an input image, (b) is a segmentation mask, and (c) is a heatmap image of abnormal area localization.

Breast cancer image segmentation is critical in labeling each pixel of an abnormal finding in X-Ray images for detecting and segmenting lesions. The segmentation combining the heatmap method of abnormal region localization in breast cancer X-Ray helps doctors identify a particular area of the disease and extract detailed information for a more accurate diagnosis.

5 Conclusions

This paper presents a method for segmentation of the abnormal regions in breast cancer X-Ray images using U-Net. The national cancer hospital radiologists have developed the image dataset of Normal, Benign, and Malignant classes. We use the breast cancer X-Ray image dataset provided by the national cancer hospital's radiologists to train the U-Net model for the segmentation of Benign and Malignant masses. For the breast cancer area segmentation task, the model quality is evaluated by comparing segmentation masks of model prediction with doctors' annotation masks on the pixel level. A Dice score metric is selected to evaluate segmentation models. Our model achieves a 0.743457 Dice score on the test set. The result shows that the segmentation model has performed well. Finally, to support more information in visualizing the Benign and Malignant regions in breast cancer X-Ray images, we combine the segmentation and the heatmap methods for abnormal region localization in breast cancer X-Ray. It helps radiologists identify a particular area of breast cancer images.

Acknowledgements This work is partly supported by the Grant of the Project KC-4.0-18/19-25 "Research on the development of an intelligent system assisted in breast cancer diagnosis based on breast cancer X-Ray images and pathology images". We thank Prof. Anh Nguyen from Auburn University, AL, USA, for his advice on this work.

References

1. N.H. Phuong, H.M. Toan, L.T. Linh, N.N. Cuong, B.M. Hanh, Weakly supervised localization of the cancer regions in breast cancer X-Ray images, in *Biomedical and Other Applications of Soft Computing*, ed. By N.H. Phuong, V. Kreinovich (Springer, Cham, Switzerland), to appear
2. J. Howard, S. Gugger, Fastai: a layered API for deep learning. Information **11**, 108 (2020). https://doi.org/10.3390/info11020108)
3. W. Shi, J. Caballero, F. Huszár, J. Totz, A.P. Aitken, R. Bishop, Z. Wang, Real-time single image and video super-resolution using an efficient sub-pixel convolutional neural network, in *Proceedings of the IEEE Conference on Computer Vision and Pattern Recognition* (2016), pp. 1874–1883
4. L.R. Dice, Measures of the amount of ecologic association between species. Ecology **26**(3), 297–302 (1945)
5. P. Jaccard, The distribution of the flora in the alpine zone 1. New Phytologist. **11**(2), 37–50 (1912)
6. T.T. Tanimoto, An elementary mathematical theory of classification and prediction. Internal IBM Technical Report (1958)
7. https://docs.opencv.org/4.x/d5/daf/tutorial_py_histogram_equalization.html
8. G. Cao, L. Huang, H. Tian, X. Huang, Y. Wang, R. Zhi, Contrast enhancement of brightness-distorted images by improved adaptive gamma correction. Comput. Electr. Eng. **66**, 569–582 (2018)
9. V. Yeghiazaryan, I. D. Voiculescu, Family of boundary overlap metrics for the evaluation of medical image segmentation. J. Med. Imaging **5**(1), 015006 (2018)
10. S. Vargas, M.I. Vera, M. Vera, J. Salazar-Torres, Y. Huérfano, O. Valbuena, E. Gelvez-Almeida, Space-occupying lesions identification in mammary glands using a mixed computational strategy. J. Phys.: Conf. Ser. **1414**(1), 012016 (2019), IOP Publishing
11. S.S. Ittannavar, R.H. Havaldar, Segmentation of breast masses in mammogram image using multilevel multiobjective electromagnetism-like optimization algorithm. BioMed Res. Int. (2022)
12. A. Anaya-Isaza, L. Mera-Jiménez, J.M. Cabrera-Chavarro, L. Guachi-Guachi, D. Peluffo-Ordóñez, J.I. Rios-Patiño, Comparison of current deep convolutional neural networks for the segmentation of breast masses in mammograms. IEEE Access **9**, 152206–152225 (2021)
13. J. Long, E. Shelhamer, T. Darrell, Fully convolutional networks for semantic segmentation, in *Proceedings of The IEEE Conference on Computer Vision and Pattern Recognition*, (2015), pp. 3431–3440
14. W. Liu, A. Rabinovich, A.C. Berg, Parsenet: looking wider to see better (2015) arXiv:1506.04579
15. G. Wang, W. Li, S. Ourselin, T. Vercauteren, Automatic brain tumor segmentation using cascaded anisotropic convolutional neural networks, in *International MICCAI Brainlesion Workshop* (Springer, Berlin, 2017), pp. 178–190
16. Y. Yuan, M. Chao, Y.C. Lo, Automatic skin lesion segmentation using deep fully convolutional networks with jaccard distance. IEEE Trans. Med. Imaging **36**(9), 1876–1886 (2017)
17. N. Liu, H. Li, M. Zhang, J. Liu, Z. Sun, T. Tan, Accurate iris segmentation in non-cooperative environments using fully convolutional networks, in *2016 International Conference on Biometrics (ICB)* (IEEE, 2016), pp. 1–8

18. O. Ronneberger, P. Fischer, T. Brox, U-net: Convolutional networks for biomedical image segmentation, in *International Conference on Medical Image Computing and Computer-assisted Intervention* (Springer, Cham, 2015), pp. 234–241
19. F. Milletari, N. Navab, S.A. Ahmadi, V-net: fully convolutional neural networks for volumetric medical image segmentation, in *2016 Fourth International Conference on 3D Vision* (3DV) (IEEE, 2016), pp. 565–571
20. Z. Zhou, M.M. Rahman Siddiquee, N. Tajbakhsh, J. Liang, Unet++: a nested u-net architecture for medical image segmentation, in *Deep Learning in Medical Image Analysis and Multimodal Learning for Clinical Decision Support* (2018), pp. 3–11
21. Ö. Çiçek, A. Abdulkadir, S.S. Lienkamp, T. Brox, O. Ronneberger, 3D U-Net: learning dense volumetric segmentation from sparse annotation, in *International Conference on Medical Image Computing and Computer-Assisted Intervention MICCAI 2016* (2016), pp. 424–432

Why FLASH Radiotherapy is Efficient: A Possible Explanation

Julio C. Urenda, Olga Kosheleva, Vladik Kreinovich, and Nguyen Hoang Phuong

Abstract Usually, a cancer radiotherapy session lasts between 10 and 20 min. Technically, it is possible to transmit the dose faster, but traditionally, medical doctors were reluctant to do it, since they were afraid of negative effects of such a speedy treatment. Recent experiments show, however, that these fears are unfounded; moreover, transmitting the whole radiation dose in a shorter time turns out to be more beneficial for the patients. In this paper, we provide a possible geometric explanation for this empirical phenomenon.

1 Formulation of the Problem

What is radiotherapy: a brief reminder. Radiotherapy is one of the most effective ways of fighting cancer, when radiation is applied to the tumor.

Traditional radiotherapy: in brief. Traditionally, each radiotherapy session lasts between 10 and 30 min.

From the purely engineering viewpoint, it is possible to apply the same overall dose of radiation faster. However, traditionally, medical doctors were reluctant to bring this dose too fast, since they were afraid of undesirable side effects of radiation-per-hour.

J. C. Urenda · V. Kreinovich (✉)
Department of Computer Science, University of Texas at El Paso 500 W. University, El Paso, TX 79968, USA
e-mail: vladik@utep.edu

J. C. Urenda
e-mail: jcurenda@utep.edu

O. Kosheleva
Department of Teacher Education, University of Texas at El Paso 500 W. University, El Paso, TX 79968, USA
e-mail: olgak@utep.edu

N. H. Phuong
Artificial Intelligence Division, Information Technology Faculty, Thang Long University, Nghiem Xuan Yem Road, Hoang Mai District, Hanoi, Vietnam

N. H. Phuong and V. Kreinovich (eds.), *Deep Learning and Other Soft Computing Techniques*, Studies in Computational Intelligence 1097,
https://doi.org/10.1007/978-3-031-29447-1_13

FLASH radiotherapy. Recently, it turned out that not only these fears were unfounded: a very fast session, when the whole dose of radiation is sent in a few seconds, seem to work even better than the traditional radiotherapy.

Specifically, experiments on animals have shown that such *FLASH* radiotherapy actually decreases the undesirable effects on the surrounding healthy cells; see, e.g., [1, 2] and references therein. These results make FLASH radiotherapy a promising approach to cancer treatment.

Problem. How can we explain these unexpected experimental results?

What we do in this paper. In this paper, we provide a possible explanation for the success of FLASH radiotherapy.

2 Our Explanation

Let us formulate the problem in precise terms. The main objective of a radiotherapy session is to transmit the overall radiation dose D_0. How we do it can be characterized by a function $D(t)$ that describes how much radiation has been transmitted during this session by time t (starting from the beginning of this section).

We start at time $t = 0$, when the dose-so-far is 0: $D(0) = 0$. At each moment of time, we can only add more radiation, so this function is (non-strictly) increasing: if $t \le t'$ then $D(t) \le D(t')$. At the end of the session, we should have $D(t) = D_0$–and this amount should remain the same until the next session.

Since the time until the next session–usually, several weeks–is much larger than the session's duration, it makes sense, when describing the current session, to ignore this future session—which is too far away from now—and to simply assume that the function $D(t)$ remains equal to D_0 for all non-negative values t.

In these terms, selecting an appropriate schedule means selecting a (non-strictly) increasing function $D(t)$ for which $D(0) = 0$ and

$$\lim_{t \to \infty} D(t) = D_0.$$

There are many such functions, which of them should we choose? Informally, we should select the best of these functions, the question is how we describe "the best" in precise terms.

How to describe "the best" in precise terms: general case. Usually, "the best" means that some objective function attains the largest (or the smallest) possible value. However, this is not the most general way of describing optimality. For example, if you have two different doses that have the same curing effect, it is reasonable to use this non-uniqueness to optimize something else: e.g., minimize the possibility of negative side effects. In this case, instead of the original single objective function

$f(a)$, we have a more complicated scheme, when an alternative a is better than an alternative b if either $f(a) > f(b)$, or if $f(a) = f(b)$ and $g(a) < g(b)$ for some other function $g(a)$.

If this more complex scheme still selects several alternatives, we can use this non-uniqueness to optimize something else, etc., until we reach the final optimality criterion in which we have only one optimal alternative.

The only thing we can say about such more general optimization settings is that we should be able, for any two alternatives a and b, to decide whether a is better than b (we will denote it by $a > b$), or b better than a ($b > a$), or a and b have the same quality (we will denote it by $a \sim b$). These relations $a > b$ and $a \sim b$ should satisfy natural consistency requirements: e.g., if a is better than b and b is better than c, then a should be better than c. Thus, we arrive at the following definition.

Definition 1 Let *A be a set. Its elements will be called alternatives.*

- By an *optimality criterion*, we mean a pair of binary relations $\langle >, \sim \rangle$ that satisfy the following conditions for all a, b, and c:
  - if $a > b$ and $b > c$, then $a > c$;
  - if $a > b$ and $b \sim c$, then $a > c$;
  - if $a \sim b$ and $b > c$, then $a > c$;
  - if $a \sim b$ and $b \sim c$, then $a \sim c$;
  - if $a > b$, then we cannot have $a \sim b$.
- We say that an alternative a_{opt} is *optimal* with respect to the optimality criterion $\langle >, \sim \rangle$ if for every $a \in A$, we have either $a_{\text{opt}} > a$ or $a_{\text{opt}} \sim a$.
- We say that the optimality criterion is *final* if there exists exactly one alternative which is optimal with respect to this criterion.

What are alternatives in our case. In our case, alternatives are different (non-strictly) increasing functions $D(t)$ for which $D(0) = 0$ and $D(t) \to D_0$ as $t \to \infty$.

Definition 2 Let D_0 be a constant. By a *D_0-alternative*, we mean a (non-strictly) increasing function $D(t)$ for which $D(0) = 0$ and $D(t) \to D_0$ as $t \to \infty$.

Natural invariance. There is no fixed unit of time relevant for this process, so it makes sense to require that the optimality criterion will not change if we use a different measuring unit to measure time.

If we know the dependence $D(t)$ in the original scale, how will this dependence look like in the new scale? If we replace the original measuring unit by a one which is λ times larger, then moment t in the new scale corresponds to moment $\lambda \cdot t$ in the original scale. For example, if we replace second with minutes—which are 60 times larger, then 2 minutes in the new scale is equivalent to $2 \cdot 60 = 120$ seconds.

In general, the value $D_{\text{new}}(t)$ corresponding to moment t in the new scale is thus equal to the value $D(\lambda \cdot t)$ when time is described in the original scale. Thus, $D_{\text{new}}(t) = D(\lambda \cdot t)$, and we arrive at the following definition.

Definition 3 Let D_0 be a real number.

- For every $\lambda > 0$ and for every D_0-alternative $D(t)$, by a λ-*rescaling* $R_\lambda(D)$, we mean a D_0-alternative $D_{\text{new}}(t) \stackrel{\text{def}}{=} D(\lambda \cdot t)$.
- We say that the optimality criterion of the set of all D_0-alternatives is *scale-invariant* if for every $\lambda > 0$ and for every two D_0-alternatives a and b, we have the following:
  - if $a > b$, then $R_\lambda(a) > R_\lambda(b)$, and
  - if $a \sim b$, then $R_\lambda(a) \sim R_\lambda(b)$.

Main result. Now, we are ready to formulate our main result.

Proposition. *Let D_0 be a real number, and let $(<, \sim)$ be a final scale-invariant optimality criterion on the set of all D_0-alternatives. Then, the optimal D_0-alternative has the form $D(t) = D_0$ for all $t > 0$.*

Discussion. This result explains the empirical fact that an instantaneous ("flash") radiotherapy indeed leads to the best medical results.

Proof 1°. Let us first prove that for every final scale-invariant optimality criterion on the set of all D_0-alternatives, the optimal D_0-alternative D_{opt} is itself scale-invariant, i.e., $R_\lambda(D_{\text{opt}}) = D_{\text{opt}}$ for all $\lambda > 0$.

Indeed, by definition, the fact that D_{opt} is optimal means that for every D_0-alternative D, we have either $D_{\text{opt}} > D$ or $D_{\text{opt}} \sim D$. This is true for every D_0-alternative D, thus, this property holds for $R_{\lambda^{-1}}(D)$, i.e., we have either $D_{\text{opt}} > R_{\lambda^{-1}}(D)$ or $D_{\text{opt}} \sim R_{\lambda^{-1}}(D)$.

Since the optimality criterion is scale-invariant, we can conclude that either $R_\lambda(D_{\text{opt}}) > R_\lambda(R_{\lambda^{-1}}(D)) = D$ or $R_\lambda(D_{\text{opt}}) > R_\lambda(R_{\lambda^{-1}}(D)) = D$. This is true for all D_0-alternatives D. Thus, by definition of optimality, this means that the D_0-alternative $R_\lambda(D_{\text{opt}})$ is also optimal.

However, we assumed that our optimality criterion is final. This means that there is only one optimal D_0-alternative, and thus, $R_\lambda(D_{\text{opt}}) = D_{\text{opt}}$. The statement is proven.

2°. Let us now use the result from Part 1 of this proof to prove the Proposition, i.e., to prove that the optimal D_0-alternative has the desired flash form.

Indeed, the equality $R_\lambda(D_{\text{opt}}) = D_{\text{opt}}$ means that the values of these two functions coincide for all t. By definition of λ-rescaling, this means that for every t and every $\lambda > 0$, we have $D_{\text{opt}}(\lambda \cdot t) = D_{\text{opt}}(t)$. In particular, by taking $\lambda = s > 0$ and $t = 1$, we conclude that for every $s > 0$, we have $D_{\text{opt}}(s) = D_{\text{opt}}(1)$. Thus, the function $D_{\text{opt}}(s)$ attains the same constant value $D_{\text{opt}}(1)$ for all $s > 0$.

In particular, for $s \to \infty$, we have $D_{\text{opt}}(s) \to D_{\text{opt}}(1)$. By definition of a D_0-alternative, this limit must be equal to D_0. Thus, $D_{\text{opt}}(1) = D_0$ and therefore, for all $s > 0$, we have $D_{\text{opt}}(s) = D_0$.

The Proposition is proven.

Acknowledgements This work was supported in part by the National Science Foundation grants 1623190 (A Model of Change for Preparing a New Generation for Professional Practice in Computer Science), and HRD-1834620 and HRD-2034030 (CAHSI Includes), and by the AT&T Fellowship in Information Technology. It was also supported by the program of the development of the Scientific-Educational Mathematical Center of Volga Federal District No. 075-02-2020-1478, and by a grant from the Hungarian National Research, Development and Innovation Office (NRDI).

References

1. B. Lin, F. Gao, Y. Yang, D. Wu, Y. Zhang, G. Feng, T. Dai, X. Du, FLASH radiotherapy: history and future, in *Frontiers in Oncology*, vol. 11, Paper 644400 (2021)
2. P. Montay-Gruel, M.M. Acharya, O. Gonçalves Jorge, B. Petit, I.G. Petridis, P. Fuchs, R. Leavitt, K. Petersson, M. Gondré, J. Ollivier, R. Moeckli, F. Bochud, C. Bailat, J. Bourhis, J.F. Germond, C.L. Limoli, M.C. Vozenin, Hypofractionated FLASH-RT as an effective treatment against Glioblastoma that reduces neurocognitive side effects in mice. Clin. Cancer Res. **27**(3), 775–784 (2021)

Data Augmentation Techniques Evaluation on Ultrasound Images for Breast Tumor Segmentation Tasks

Trang Minh Vo, Thien Thanh Vo, Tan Tai Phan, Hai Thanh Nguyen, and Dien Thanh Tran

Abstract A periodic health check with physical examination and assessment of the current health status is among the best approaches to protect your health. Through that, we can know if we have any health problems and promptly offer a treatment plan if the problems are detected. In most cases, early diagnosis and disease detection are very important, especially for cancer. Breast cancer is serious and also a very common disease in women. However, if the disease is detected early, the chance of a cure is very high. Deep learning-based segmentation methods have been introduced to detect Breast cancer tumors, but we are facing challenges in the limitations of data. Although some data augmentation approaches have been presented, the number of augmented samples should be further considered. This work has examined the efficiency of data augmentation techniques by Brightness and Rotation with various ratios of increased samples on ultrasound images. Data augmentation improves the performance using U-NET to perform segmentation tasks for breast cancer diagnosis. The experimental results show that the rotation technique can increase the average performance in the training and test phases.

1 Introduction

Ultrasound is important in disease diagnosis and treatment and is a safe, non-invasive method. In many cases, ultrasound is the preferred choice, supporting doctors in screening and accurate diagnosis. Unfortunately, cancer is currently one of the leading global health concerns. Overall, cancer incidence and mortality are increasing

T. M. Vo · T. T. Vo · T. T. Phan · H. T. Nguyen (✉) · D. T. Tran
Can Tho University, Can Tho, Vietnam
e-mail: nthai.cit@ctu.edu.vn

T. T. Phan
e-mail: pttai@cit.ctu.edu.vn

D. T. Tran
e-mail: ttdien@ctu.edu.vn

N. H. Phuong and V. Kreinovich (eds.), *Deep Learning and Other Soft Computing Techniques*, Studies in Computational Intelligence 1097,
https://doi.org/10.1007/978-3-031-29447-1_14

rapidly around the world. Currently, tumor diseases are quite common. Cancer is a dangerous disease that affects health and has a high probability of death, which needs to be detected early and treated promptly to help patients have reasonable treatment methods. Especially breast cancer is a disease many women care about and learn to screen. Early screening offers a good chance of treatment, and some cancers even have a chance of being cured. One tool that plays an important role in tumor detection is imaging. Breast cancer will become the most commonly diagnosed cancer in the world. In 2022, it is estimated that 43.250 women and 530 men will die of breast cancer.[1] Breast cancer occurs in both men and women, but breast cancer is most common in women over 50 years old. In addition, the disease can be passed from mother to child. Therefore, in addition to supplementing with adequate nutrients for the body, being active, and eating healthy, it is also necessary to have regular health check-ups to screen for diseases.

In Vietnam, breast cancer ranks first among cancers in women. The incidence of breast cancer in the world and Vietnam, in particular, has increased in recent years. Tumors in the chest can be detected through the noted symptoms, and support imaging techniques such as ultrasound, X-ray, and MRI are often used to diagnose the disease. With the advancement of modern science and technology, technology increasingly dominates all human activities, helping to optimize jobs with high accuracy, supporting people in activities, and living actively. Especially with the development of artificial intelligence, imaging tools play an increasingly important role in the screening, early diagnosis, monitoring, and treatment of diseases. The cancer cell's shape determines whether the tumor is benign or malignant. Data augmentation techniques are concerned with improving the performance of some algorithms in breast image analysis. For example, in [1], the authors used Generative Adversarial Network and convolutional neural network with crop techniques. Another study in [2] attempted Data augmentation techniques with Unet on segmentation of various organs. However, using Unet and augmentation techniques for ultrasound images for breast tumor segmentation tasks, especially for brightness, is still limited. Therefore, our study investigated some data augmentation techniques with Unet with contributions as follows.

- We investigated the effect of various data augmentation techniques when we deployed U-net network architecture for breast cancer diagnosis.
- Data augmentation on the breast ultrasound image data set, with brightness and rotation transformations, increases the comparative data of the generated images. The results show that with data augmentation by rotation, the model can be stable and have higher performance and less over-fitting than the other.
- We also try different scaling of images using data augmentation techniques. As observed from the results, the performance is improved on testing sets. However, with a rate increase of 25%, the performance increases gradually from the testing set and increases significantly compared to the percentage performance increase of 50%. In the original data set, 780 samples average results Testing = 89%(0.890).

[1] https://www.stopbreastcancer.org/information-center/facts-figures/.

After data augmentation, 936 samples and 1092 samples average testing results of 89.6% (0.896) and 94% (0.940), respectively.

The rest of the paper is organized as follows. First, we will discuss several related studies on medical image analysis methods in Sect. 2. After that, Sect. 3 elaborates on our architecture and algorithms. Subsequently, Sect. 4 will explain our experiments for the system. Finally, we will summarize our studies' key features and development directions (Sect. 5).

2 Related Work

NecipCinar, AlperOzcan, MehmetKaya [3] proposed A hybrid Dense-Net121 U-Net model for brain tumor segmentation from MR Images. Recently, researchers proposed different MRI techniques to detect brain tumors with the possibility of uploading and visualizing the image, aiming to segment brain tumors using deep-learning MR images. The U-Net architecture, one of the deep learning networks, is used as a hybrid model with pre-trained Dense-Net121 architecture for the segmentation process. During training and testing of the model, we focus on smaller sub-regions of tumors that comprise the complex structure. The experimental results indicate that our model performs better than other state-of-the-art methods presented in this particular area. Specifically, the best Dice Similarity Coefficient (DSC) is obtained by using the proposed approach to segment whole tumor (WT), core tumor (CT), and enhancing tumor (ET). The work in [4] provided expert-level prenatal detection of complex congenital heart disease. Congenital heart disease is the most common congenital disability. Fetal screening ultrasound provides five heart views that can detect 90% of complex Congenital heart diseases. Using 107,823 images from 1,326 retrospective echo cardiograms and screening ultrasounds from 18 to 24 weeks fetuses, a neural network was created to identify recommended cardiac imaging and differentiate between normal hearts and complex hearts.

The work in [5] achieved remarkable results in classifying and segmenting natural images and computer vision tasks. However, medical images face challenges due to a need for ground truth for segmentation tasks. As a result, this study proposed and developed automatic methods for segmenting breast lesions from ultrasound images with few label images. The proposed U-NET model employs the GELU activation function and has been tested on 620 breast ultrasound images, including 310 benign and 310 malignant cases. The quantitative analysis includes accuracy, loss function, dice similarity coefficient, and precision. The qualitative analysis compares original images, masks, image predicted, and the output result, mask predicted binary image. The proposed U-NET model outperformed previous methods, achieving a DSC of 99.62%, a loss of 3.01%, an accuracy of 98.15%, and a precision of 91.82%. In addition, the authors have identified some of the innovations made to the original U-NET model. The Authors in [6] proposed a fully automatic segmentation algorithm consisting of a fuzzy, fully convolutional network and accurately fine-tuning post-processing based on breast anatomy constraints. In the first part, the image is pre-processed by contrast enhancement, and wavelet features are employed for image augmentation. Then, a fuzzy membership function transforms the augmented BUS images into the fuzzy domain. The elements from convolutional layers are processed using fuzzy logic as well. The conditional random fields post-process the segmentation result. Finally, the location relation among the breast anatomy layers is utilized to improve the performance. The proposed method is applied to the data set with 325 BUS images and achieves state-of-the-art performance compared to existing processes with a true positive rate of 90.33%, false positive rate of 9.00%, and intersection over union (IOU) 81.29% on tumor category.

Li et al. [7] proposed a two-step deep learning framework for breast tumor segmentation in breast ultrasound (BUS) images which requires only a few manual labels. Wang et al. [8] proposed a segmentation framework combining an active contour module and deep learning adversarial mechanism to segment breast tumor lesions. The Deformed U-Net performs pixel-level segmentation for breast ultrasound images. The active contour module refines the tumor lesion edges, and the refined result provides loss information for Deformed U-Net. Therefore, the Deformed U-Net can better classify the edge pixels. The proposed method for segmenting the tumor lesions in breast ultrasound image obtains dice coefficient: 89.7%, accuracy: 98.1%, precision: 86.3%, mean-intersection-over-union: 82.2%, recall: 94.7%, specificity: 98.5% and F1score: 89.7%.

The authors in [9] proposed an image segmentation method on breast tumor ultrasound images based on the U-Net framework, combined with residual block and attention mechanism. The experimental results show that the proposed method can learn the analysis and processing of actual complex breast tumor ultrasound images as reliable medical diagnosis assistance for medical staff and that the Dice index value of the proposed method can reach 0.921. The work in [10] there has been an innovation in the proposed expanded training approach to obtain an expanded U-Net. In the study in [11], proper treatment for the patients and symptoms should be observed properly, and an automatic prediction system is needed to classify the tumor as benign or malignant. As a general convolutional neural network, its role focuses on classifying images, where input is an image and output is one label. Biomedical cases enable us to discern whether a disease occurs and locate the abnormality. U-Net is devoted to solving this problem. This research has proposed a U-NET-based architecture for segmenting tumor regions in histopathological images. The proposed method gives an overall accuracy of 94.2 with fewer data sets. A survey in [12] analyzed a vast of data augmentation on mammogram images.

3 Methods

This study evaluates the effects of data augmentation with various image sizes on the segmentation tasks on breast ultrasound images, as shown in Fig. 1. First, from original images with the resolution of 500×500, we downsize the original images

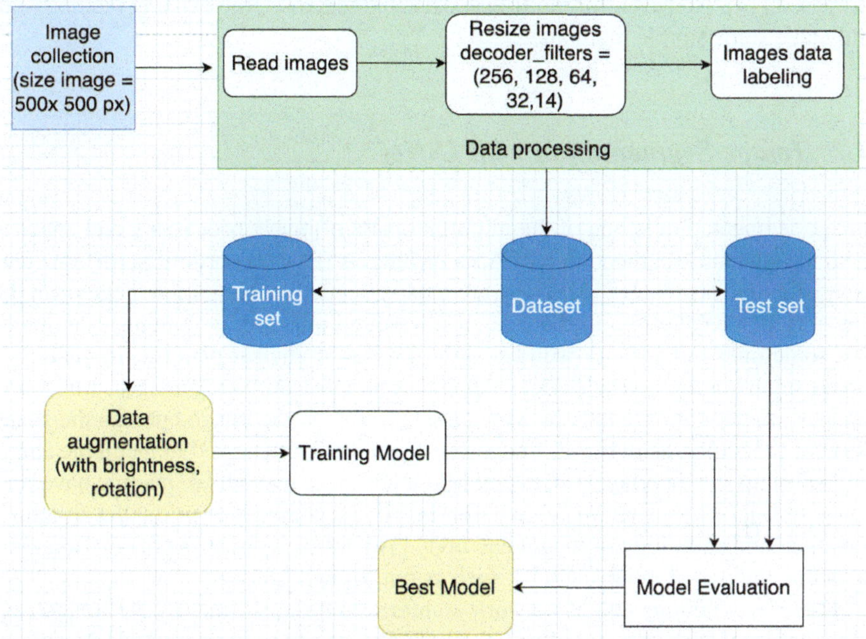

Fig. 1 The overall workflow for the proposed method

to 192 × 192. Then, we separately perform the augmentation techniques, including brightness and rotation on the training set, before fetching into segmentation models to compare the efficiency of the two techniques. The details are presented as follows.

3.1 Data Augmentation

A major limitation of supervised Deep Neural networks in medical imaging is the need for large annotated data sets. Though quite efficient in enhancing the performance of deep learning networks, current data augmentation methods do not include complex transformations. The extracted data variability was then transferred through data augmentation to a small database to train a deep learning-based segmentation algorithm. Significant improvements are observed compared to usual data augmentation techniques [13]

Data augmentation is a commonly used method for training networks in deep learning. For object detection tasks, data augmentation methods include random flip, random crop, etc. For semantic segmentation tasks, data enhancement includes brightness, zooming in/out, flipping the image, rotating, random saturation transformation, etc., and many multi-task networks based on deep learning have been proposed [14]. However, for multi-task data augmentation, the existing methods can only take some simple methods.

3.2 *Image Segmentation with U-Net*

Image segmentation is one of the important parts of image processing and analysis with the standard segmentation of Area of Interest (ROI) according to certain characteristics and attributes [15]. There are many effective segmentation methods in the category of image tools. In the case of ultrasound imaging, communication methods sometimes fail to work well because of noise or unclear patterns. In medicine, image segmentation can help doctors diagnose tumors from X-ray images, ultrasound images, etc.; it not only tells us where the tumor is on the image but also the shape form of the tumor, as in Fig. 2. Image segmentation is often used in medicine and is an important step in image processing Major advances in medical imaging have provided physicians with powerful and non-invasive techniques to probe the structure, function, and pathology of the human body [16]. Medical image segmentation aims to define regions of interest (ROI), such as tumors and lesions.

U-Net is a famous neural network architecture designed mainly for image segmentation and outputs a label for each pixel. The structural basis of the U-Net architecture consists of two paths. The first path is the path that gets the encoding or parsing of the path, the same as a parsing network normal and supply type information. The second is called a decoder or path synthesizer, which includes switches and connections to features from the encoder. This width allows the network to search for the type of information. In addition, the path extension also increases

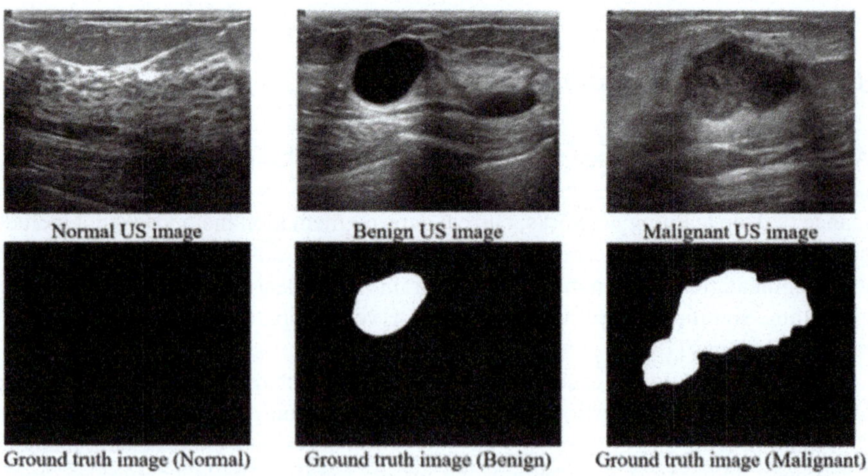

Fig. 2 Examples of the original ultrasound images and their masks

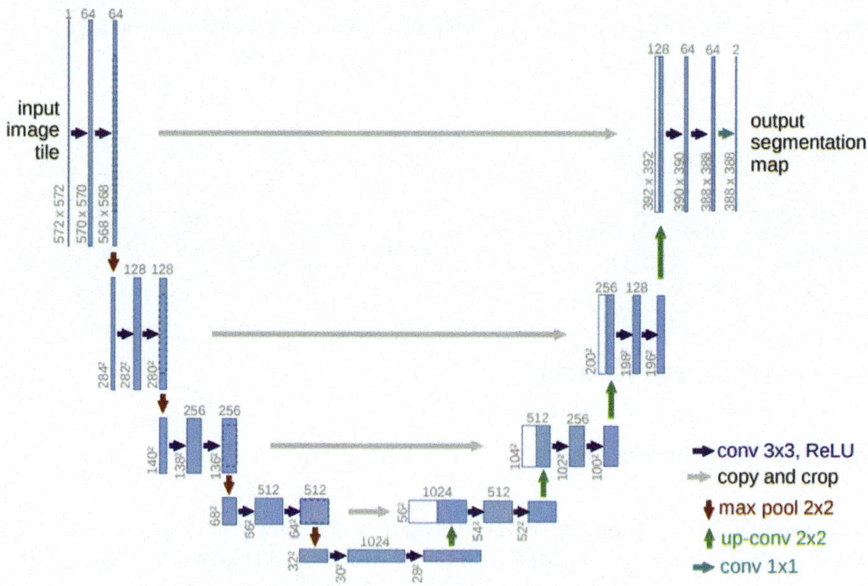

Fig. 3 An illustration of U-Net network architecture [17]

the resolution of the output, which can then be moved to the bottom of the layer display to produce a full segment image. The result is almost symmetrical, giving it a U-like shape. The implementation by most syndicates is to segment the entire image into a single label. U-net is a start-up design that aims to segment medical images, and segments are widely used in medicine the workflow as detailed in Fig. 3, for Example, X-Ray images, MRI images, ultrasound images [17]. Furthermore, the U-net network applies end-to-end training to medical image analysis. As one of CNN's most important semantic segmentation frameworks, U-net plays a very important role in medical image analysis. The main idea is that the image with a fixed size is dimension reduced to make it conform to the size of the display area and generate thumbnails of the corresponding image to extract deeper image features. Then, up-sampling is used to enlarge the image, and each layer feature of down-sampling and up-sampling is fused via copy and crop. Finally, an important advantage of the U-net network is that it has many feature channels in the up-sampling part allowing the network to propagate contextual information to higher-resolution layers [18].

Table 1 The original data set and data sets with augmented images with different techniques

	Original data	25% augmented samples	Augmented 50% samples
Total samples	624	780	936
The original training set	780	936	1092
The increased images	0	156	312
Increased ratio	0	0.25	0.5

4 Experimental Results

4.1 Environmental Setting

Our research is based on the Python programming language using open source libraries such as TensorFlow along with Keras, Numpy, and Open CV-python, hyperparameters including a batch size of 16, the numbers of filters of 256, 128, 64, 32, 16 with the filter size of 3×3 corresponding the first, second, third, fourth, fifth layers, respectively, running to 100 epochs. Algorithms are run on Py-Charm, an integrated development platform specially designed for python to support running Python code directly through a web browser, suitable for machine learning research, data analysis materials, and education.

4.2 Data Description

The data set collected in 2018 included 780 breast ultrasound images of 600 women aged 25–75. The average image size is 500×500 pixels, png format, consisting of 133 normal, 487 benign, and 210 malignant samples obtained from the NCBI (National Center for Biotechnology Information) database [19]. In addition, we augmented samples from the training set with various ratios using brightness and rotation (Table 1).

4.3 Evaluation Metrics

The data set was split using 5-fold-cross validation with training sets approximately occupying 80% and the remaining 20% for the test set. The performance is evaluated by Intersection Over Union (IOU), which is the ratio of the area of overlap to the area of union. It is a measure used to evaluate algorithms and models for object recognition on data sets, also known as Jaccard's index, calculating the "Area of Intersection" of

Table 2 Training and testing results in IOU on the data set with 25% of samples increased (corresponding to 156 samples) using brightness and rotation techniques

Fold	Dataset 936 samples (Brightness)		Dataset 936 samples (Rotation)	
	Training	Testing	Training	Testing
1	0.689	0.692	0.801	0.748
2	0.977	0.873	0.978	0.887
3	0.962	0.914	0.981	0.938
4	0.984	0.961	0.985	0.929
5	0.987	0.978	0.986	0.975
Average	0.920	0.884	0.947	0.896

two boxes over the "Area of Union" of the same two boxes. During the decades, many studies have been conducted to improve efficiency and robustness in detecting and segmenting breast tumors based on size, shape, location, and contrasts. In addition, the effectiveness of group normalization with attention gate is also explored with skip connections to segment small-scale breast tumors using several highlighted salient features. Therefore, the larger the IOU, the better, which means that the intersection is large and the component is small (the predicted label is the same as the real label) [20].

4.4 *Scenario 1: Various Data Augmentation Techniques Comparison (Brightness and Rotation)*

From the results in Table 2, the performance in IOU of training and testing after increasing the data from the original data set of 780 samples to 936, the training and testing in IOU are based on data augmentation techniques by brightness and by rotation. The results show that generated images by rotation give a high segmentation accuracy than brightness. We noticed that training with augmented images by brightness could cause over-fitting. The training performance with brightness is higher than the rotation approach, but the test set's performance is reversed.

The segmentation accuracy increases steadily when further expanding the data set by the Rotation method from 780 samples to 936 and 1092 samples.

4.5 *Scenario 2: The Comparison of the Number of Augmented Images*

Next comes is training and testing result in the Data Augmentation method by rotation based on IOU, including training and testing on all three data sets, including the

Table 3 Training and testing performance in IOU of D1, D2 and D3 using data augmentation by rotation

Fold	Dataset 780 samples		Dataset 936 samples		Dataset 1092 samples	
	Training	Testing	Training	Testing	Training	Testing
1	0.940	0.776	0.801	0.748	0.978	0.941
2	0.977	0.897	0.978	0.887	0.941	0.902
3	0.963	0.909	0.981	0.938	0.938	0.925
4	0.985	0.961	0.985	0.929	0.984	0.966
5	0.921	0.905	0.986	0.975	0.985	0.965
Average	0.958	0.890	0.947	0.896	0.965	0.940

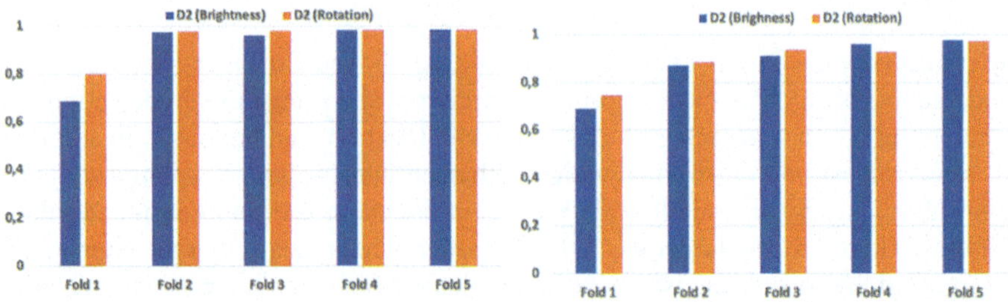

Fig. 4 Training and testing results comparison in IOU of D2 using brightness and rotation for data augmentation

original data set with 780 samples (D1), a data set including 936 samples augmented with a rate of 25% (D2), and a data set of 1092 samples with an augmented ratio of 50% (D3) (Table 3).

The performance on D2 using rotation is greater than the brightness technique to prove that model segmentation is highly efficient. For example, the IOU average of Testing with Data set 936 samples (data augmentation by brightness) = 88.4% < Data set 936 samples (data augmentation by rotation) = 89.6% should model segmentation with data augmentation by rotation is more accurate than by rotation.

Performance comparison of U-net performing on the original data set and the data set with data augmentation using rotation is revealed in Fig. 4. The performance of U-Net on D1, D2, and D3 is 0.890, 0.896, and 0.940, respectively. As observed, data augmentation can improve the accuracy of segmentation. The augmented ratio of 25% can increase the performance by 0.006, while a ratio of 50% can enhance the performance by 0.044. However, the difference between the ratios of 25 and 50% increases significantly (Fig. 5).

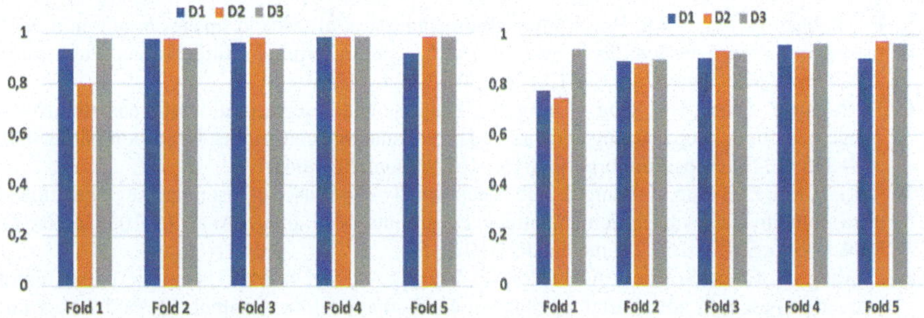

Fig. 5 Training and testing results comparison in IOU for various augmented ratios

5 Conclusion

In this study, we have evaluated data augmentation techniques of brightness and rotation to compare their effects. With the original dataset, we got an IoU of 89%, while increasing the number of samples by 25% can slightly improve the performance with 89.6%. We noticed that the result could be improved to 94% with a data increase ratio of 50%. As observed, when increasing data for training, the accuracy of partitioning results also increases. At the same time, comparing data augmentation methods by brightness and rotation, the results show that the rotation method is better than brightness. However, there is still a limitation in the study, such as not yet building optimized models with even larger data sets with higher accuracy.

The study can be useful in breast tumor image segmentation, supporting medical imaging for quick and accurate results. In the future, we plan to study more methods and apply more advanced models to improve the efficiency of tumor imaging through ultrasound images.

References

1. V.K. Singh, H.A. Rashwan, S. Romani, F. Akram, N. Pandey, M.M.K. Sarker, A. Saleh, M. Arenas, M. Arquez, D. Puig, J. Torrents-Barrena, Breast tumor segmentation and shape classification in mammograms using generative adversarial and convolutional neural network. Exp. Syst. Appl. **139**, 112855 (2020). https://doi.org/10.1016/j.eswa.2019.112855
2. T. Nemoto, N. Futakami, E. Kunieda, M. Yagi, A. Takeda, T. Akiba, E. Mutu, N. Shigematsu, Effects of sample size and data augmentation on u-net-based automatic segmentation of various organs. Radiol. Phys. Technol. **14**(3), 318–327 (2021). https://doi.org/10.1007/s12194-021-00630-6
3. N. Cinar, A. Ozcan, M. Kaya, A hybrid DenseNet121-UNet model for brain tumor segmentation from MR images. Biomed. Signal Process. Control **76**, 103647 (2022). https://doi.org/10.1016%2Fj.bspc.2022.103647
4. R. Arnaout, L. Curran, Y. Zhao, J.C. Levine, E. Chinn, A.J. Moon-Grady, An ensemble of neural networks provides expert-level prenatal detection of complex congenital heart disease. Nat. Med. **27**(5), 882–891 (2021). https://doi.org/10.1038%2Fs41591-021-01342-5

5. E. Michael, H. Ma, S. Qi, Breast tumor segmentation in ultrasound images based on u-NET model, in *Advances in Intelligent Systems and Computing* (Springer International Publishing, 2022), pp. 22–31. https://doi.org/10.1007/978-3-031-14054-9_3
6. K. Huang, Y. Zhang, H. Cheng, P. Xing, B. Zhang, Semantic segmentation of breast ultrasound image with fuzzy deep learning network and breast anatomy constraints. Neurocomputing **450**, 319–335 (2021). https://doi.org/10.1016%2Fj.neucom.2021.04.012
7. Y. Li, Y. Liu, L. Huang, Z. Wang, J. Luo, Deep weakly-supervised breast tumor segmentation in ultrasound images with explicit anatomical constraints. Med. Image Anal. **76**, 102315 (2022). https://doi.org/10.1016%2Fj.media.2021.102315
8. J. Wang, G. Chen, S. Chen, A.N.J. Raj, Z. Zhuang, L. Xie, S. Ma, Ultrasonic breast tumor extraction based on adversarial mechanism and active contour. Comput. Methods Programs Biomed. **225**, 107052 (2022). https://doi.org/10.1016/j.cmpb.2022.107052
9. T. Zhao, H. Dai, Breast tumor ultrasound image segmentation method based on improved residual u-net network. Comput. Intell. Neurosci. **2022**, 1–9 (2022). https://doi.org/10.1155%2F2022%2F3905998
10. Y. Guo, X. Duan, C. Wang, H. Guo, Segmentation and recognition of breast ultrasound images based on an expanded u-net. PLOS ONE **16**(6), e0253202 (2021). https://doi.org/10.1371%2Fjournal.pone.0253202
11. M. Robin, J. John, A. Ravikumar, Breast tumor segmentation using u-NET, in *2021 5th International Conference on Computing Methodologies and Communication (ICCMC)* (IEEE, 2021). https://doi.org/10.1109/iccmc51019.2021.9418447
12. P. Oza, P. Sharma, S. Patel, F. Adedoyin, A. Bruno, Image augmentation techniques for mammogram analysis. J. Imaging **8**(5), 141 (2022). https://doi.org/10.3390/jimaging8050141
13. L. Caselles, C. Jailin, S. Muller, Data augmentation for breast cancer mass segmentation, in *Lecture Notes in Electrical Engineering* (Springer Singapore, 2021), pp. 228–237. https://doi.org/10.1007/978-981-16-3880-0_24
14. P. Huang, Y. Zhu, Multi-task data augmentation method joint object detection and semantic segmentation, in *2022 International Conference on Machine Learning and Knowledge Engineering (MLKE)* (IEEE, 2022). https://doi.org/10.1109%2Fmlke55170.2022.00032
15. H. Cheng, J. Shan, W. Ju, Y. Guo, L. Zhang, Automated breast cancer detection and classification using ultrasound images: a survey. Pattern Recognit. **43**(1), 299–317 (2010). https://doi.org/10.1016%2Fj.patcog.2009.05.012
16. D. Jayadevappa, S. Kumar, D. Murty, Medical image segmentation algorithms using deformable models: a review. IETE Techn. Rev. **28**(3), 248 (2011). https://doi.org/10.4103/0256-4602.81244
17. O. Ronneberger, P. Fischer, T. Brox, U-net: convolutional networks for biomedical image segmentation, in *Lecture Notes in Computer Science* (Springer International Publishing, 2015), pp. 234–241. https://doi.org/10.1007%2F978-3-319-24574-4_28
18. N. Siddique, S. Paheding, C.P. Elkin, V. Devabhaktuni, U-net and its variants for medical image segmentation: a review of theory and applications. IEEE Access **9**, 82031–82057 (2021). https://doi.org/10.1109/access.2021.3086020
19. W. Al-Dhabyani, M. Gomaa, H. Khaled, A. Fahmy, Dataset of breast ultrasound images. Data in Brief **28**, 104863 (2020). https://doi.org/10.1016%2Fj.dib.2019.104863
20. Y. Alzahrani, B. Boufama, Deep learning approach for breast ultrasound image segmentation, in *2021 4th International Conference on Artificial Intelligence and Big Data (ICAIBD)* (IEEE, 2021). https://doi.org/10.1109/icaibd51990.2021.9459074

Applications to Finances

Stock Price Movement Prediction Using Text Mining and Sentiment Analysis

Nguyen Thi Huyen Chau, Le Van Kien, and Doan Trung Phong

Abstract Stock market analysis utilizing data mining has become an attractive field of research in recent years. Forecasting stock price behavior in a timely fashion would greatly facilitate the investors' appropriate decisions, improve profitability and hence decrease possible losses. This study aims at investigating intelligent techniques of deep learning and machine learning to forecast stock price movements from short news data. We collected real-world data of stock prices and short news for 1652 stock codes from 3 biggest stock markets in Vietnam. We propose a novel data augmentation technique to enrich the label dataset, which improves the prediction result by 2.58% on average. By employing the ensembling technique, we could increase furthermore the testing accuracy of our models by at least 1.84%. Lastly, experimental results strongly suggest that a 7-day analysis after posting day generally reflects better the impact of the news than a 3-day analysis does.

Keywords Stock market · Text mining · News sentiment analysis

1 Introduction

Stock investors have always been striving for better profits by analyzing market information. However, according to the "efficient market hypothesis" (EMH) [1], it is believed that the market reacts instantaneously to any given news, therefore stock prices would follow a random walk and is essentially unpredictable. Still, in [2, 3], it is shown that financial market analysists and investors often focus on planning the buy and sell strategies rather than on predicting stock prices. Given this premise, the

N. T. H. Chau (✉) · L. V. Kien · D. T. Phong
Faculty of Information Technology, Thang Long University,
Nghiem Xuan Yem Road Hoang Mai District, Hanoi, Vietnam
e-mail: huyenchau@thanglong.edu.vn

D. T. Phong
e-mail: phongdt@thanglong.edu.vn

N. H. Phuong and V. Kreinovich (eds.), *Deep Learning and Other Soft Computing Techniques*, Studies in Computational Intelligence 1097,
https://doi.org/10.1007/978-3-031-29447-1_15

interest of research has been shifted to devising methods to forecast the future stock trends instead of stock prices.

For the recent two decades, artificial intelligence technology has been developing progressively, producing outstanding achievements in emulating intelligent behaviors, many of which have been proved to be adequate or even surpass human capabilities. The stock investment area can benefit from these techniques since a possible solution for quick processing of large volumes of data is the use of automatic artificial intelligence analysis to predict stock trends in a timely fashion.

Stock behavior forecast using text mining has been becoming an attractive field of research which produces numerous investigation of stock prediction using intelligent techniques or machine learning. Some of them solved the stock prediction problem by employing such methods as fuzzy theory [4], artificial neural network [5] and Support Vector Machine (SVM) [6], which are all based on time series data. Various methods have also been proposed to stock analysis based on the different kinds of text data such as news [7], blog articles and comments [8], and tweets [9].

Of the various text mining tasks, sentiment analysis concerns with classifying textual data into positive, negative and neutral sentiment in order to determine people's attitudes towards studied entities [10]. Text preprocessing also is a vital task in text mining, of which removal of HTML tags, tokenization of sentences, noun phrasing [11], document weighting, stop-word removal, stemming/lemmatization [12], TF-IDF, and extraction of named entities are some of most common steps.

Our contribution is to propose a novel approach to predict stock trends based on short news. Other articles concerning Vietnamese stock studied financial article titles from financial media sites instead of the short news [13, 14]. Also, since these article titles may have already contained adjective words that reflex the sentiment (e.g. "Vinh Hoan (VHC): August 1/2020 revenue reached VND 500 billion, export markets simultaneously increased well"), these authors first analysed the sentiment from the data then classified the stock trends. In our work, we directly classified the price trend from the data, as in our model, the sentiments "good, bad, neutral" mirror our classified trend labels.

Our contributions are presented as follows: Section 2 describes our data collection. We explain our augmentation method to enrich the label data in Sect. 3. Section 4 talks about deep learning and machine learning model development to predict stock price trend, of which our ensembling method is elaborated in Sect. 4.4. Experimental results are examined in Sect. 5. Finally, we provide a brief discussion and perspective in Sect. 6.

2 Data Collection

Stocks are certificates issued by a joint-stock company, which are journal entries or electronic data certifying that the holder has proportionate ownership in the issuing corporation. Stocks are sold predominantly on stock exchanges.

	Stock ▾		Sector	Stock market	Amount
1	YTC	CTCP Xuất nhập khẩu Y tế Thành phố Hồ Chí Minh	Bán buôn hàng tiêu dùng	UPCoM	3,080,000
2	YEG	CTCP Tập đoàn Yeah1	Quảng cáo, quan hệ công chúng và dịch vụ liên quan	HOSE	31,279,968
3	YBM	CTCP Khoáng sản Công nghiệp Yên Bái	Sản xuất sản phẩm khoáng chất phi kim	HOSE	14,299,880

Fig. 1 Stocks

Short news in securities is news whose content is about the stock market and is conveyed in a concise and selective manner.

In our research, we need to collect two types of data: the short news and the stock prices. For the experimental validation, this research used real-world data of stock prices and short news for 1652 stock codes from 3 biggest stock markets in Vietnam, namely, Hose, HNX, and UpCoM (Fig. 1). We crawled this data from the following sources: vndirect.com.vn and tinnhanhchungkhoan.vn. We also limited the scope of our curation to all available data posted on the 2 websites as of April 23, 2022.

For collection results, we obtained 2482 files of short news data and 1652 files of stock price data, with total sizes of 12MB and 95.5MB respectively. Each short news consists of 3 attributes: posting date, stock code, and content of the short news. Each price consists of 6 attributes: date, opening price, high price of the day, low price of the day, closing price, total trading volume. This study would analyse all the properties of the news data and also examine the closing price attribute of the stock price data.

3 Data Preparation

3.1 Data Cleaning

Short news data has been preprocessed by the following steps: removal of redundant marks and accentuation, deletion of unnecessary information that was enclosed in parentheses, conversion of all uppercase letters to lowercase letters, replacement of abbreviations by complete phrases, and finally token coupling into word pairs. For the last and most significant step of token coupling, we employed the pre-trained PhoBERT models which are the state-of-the-art language models for Vietnamese. PhoBERT based its pre-training approach on RoBERTa which optimizes the BERT pre-training procedure for more robust performance [15]. PhoBERT outperforms previous monolingual and multilingual approaches, obtaining new state-of-the-art performances on four downstream Vietnamese NLP tasks of part-of-speech tagging, dependency parsing, named-entity recognition and natural language inference.

Stock	Data	Filename	News	N	N + 1	N + 2	N + 3
gsp	2021-04-26	26-4-2021.txt	công ty cổ_phần vận_tải sản_phẩm khí quốc_tế t...	14.3	14.35	14.1	12.0
dhg	2021-04-26	26-4-2021.txt	ngày 07/5 là ngày giao_dịch không hưởng quyền ...	100.0	100.00	99.9	99.8
sdg	2021-04-26	26-4-2021.txt	công ty trách_nhiệm hữu_hạn tm dv được lộc và ...	35.0	38.50	40.0	40.0

Fig. 2 Data concatenation results

3.2 Short News and Stock Prices Concatenation

After preprocessing the data, we performed the concatenation of news and price data into one DataFrame (Fig. 2).

N, $N+1$, $N+2$, $N+3$ are the closing prices of respectively the posting date, one day later, 2 days later, and 3 days later.

3.3 Data Labelling

Our objective is to predict the influence of particular short news on stock prices. News is to be deemed "good" if it might affect the stock price to rise, "bad" if it might cause the price to drop, and "neutral" if it is considered to have almost no significant impact on the price. Hence in the next stage of analysis, our datasets are labeled by taking the average N' of the closing prices of either 3 or 7 days after the posting date to compare with the closing price of the posting date N. If $N < N'$, the corresponding news would been labeled as 2, if $N > N'$, it would been labeled as 0 and in the case of $N = N'$, the news would been labeled as 1.

3.4 Label Data Augmentation

Since a lot of short news had been published such a long time ago when price data had not been stored, not all collected data could be labeled with price data. To tackle this price data missing, we first employed Vietnamese SBERT to encode the short news data to numerical vectors. We then calculated the Euclidean distance between two vectors to evaluate the similarity between a given unlabeled short news to any labeled one, from which we finally estimated the unlabeled data by the trend labels of their "nearest neighbour" short news (Fig. 3).

For any 2 vectors $P(p_1, p_2, \ldots, P_n)$ và $Q(q_1, q_2, \ldots, q_n)$, we have their Euclidean distance computed as follows:

$$d(p,q)^2 = (p_1 - q_1)^2 + (p_2 - q_2)^2 + \cdots + (p_i - q_i)^2 + \cdots + (p_n - q_n)^2$$

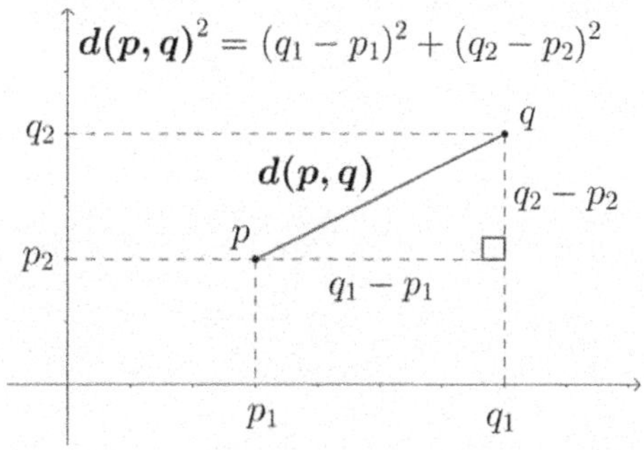

Fig. 3 The Euclidean distance of two vectors

4 Model Development

We first describe our recurrent neural network models, then our machine learning models for the stock news analysis problem.

4.1 Regressive Neural Network Model

In this subsection, we would present and evaluate the neural network approach for predicting a trend label (good, bad, neutral) for any news data. There are typically 4 types of neural networks commonly used: radial basis function neural networks (RBFNNs), feed-forward neural networks (FFNNs), convolutional neural networks (CNNs, a particular type of FFNNs), and recurrent neural networks (RNNs). Of all the four, RNNs is designed to process sequential data, have been studied and have proven to be the most apt for time series data and natural language processing. Based on these reasons, in this paper, we focused exclusively on analyzing how FFNNs can be used to predict trend labels of short news.

4.1.1 Data Encoding

For building our recurrent neural network models, the pre-trained Word2Vec [16] encoder was used to encode the short news. In this encoding model, a word and a sentence would be transformed respectively into a 1×300 dimensional vector and a 176×300 dimensional one (Fig. 4). Label data is encoded next using One Hot Encoding techniques, which is designed to work well with the Categorical Cross Entropy loss function in the model architecture that we will describe later on.

```
array([[ 0.        ,  0.        ,  0.        , ...,  0.        ,
         0.        ,  0.        ],
       [ 0.        ,  0.        ,  0.        , ...,  0.        ,
         0.        ,  0.        ],
       [ 0.        ,  0.        ,  0.        , ...,  0.        ,
         0.        ,  0.        ],
       ...,
       [ 0.00924758, -0.00312946,  0.02535261, ...,  0.0613357 ,
         0.09311175,  0.04752316],
       [ 0.04071017, -0.05782425, -0.01652207, ..., -0.00861668,
         0.04219515, -0.00249166],
       [ 0.03171563,  0.00435146, -0.0039089 , ...,  0.03654759,
        -0.10818528, -0.03179894]])
```

Fig. 4 The news vector obtained from Word2Vec encoding

4.1.2 Model Architecture

As mentioned in the previous subsection, although there are multiple deep learning methods in the literature for analysing stock news, in this paper, we focus only on recurrent neural network models in order to determine the best structure and configuration for predicting the impact of short news on stock prices. We selected some of the most widely use RNN variations, namely the RNN [17], LSTM [17], and Bidirectional Long-Short-Term Memory (BLSTM) [18, 19] architectures, for each of which we would then develop 2 models (Table 1). All models would also share the following properties:

- The input layer is of (176, 300) size
- The output layer has 3 neurons
- The activation function is Softmax
- The optimizer function is Adam
- The loss function is Categorical Crossentropy
- All models would be using bias.

Table 1 Optimal configuration for RNN, LSTM, BLSTM architectures

Model	Hidden layer	Model	Hidden layer
RNN_1	0.5 Dropout, 64 neuron RNN	LSTM_2	0.5 Dropout, 64 neuron LSTM
			0.5 Dropout, 128 neuron LSTM
RNN_2	0.5 Dropout, 64 neuron RNN	BLSTM_1	0.5 Dropout, 64 neuron BLSTM
	0.5 Dropout, 128 neuron RNN		
LSTM_1	0.5 Dropout, 64 neuron LSTM	BLSTM_2	0.5 Dropout, 64 neuron BLSTM
			0.5 Dropout, 128 neuron BLSTM

4.2 *Machine Learning Models*

Though deep learning has attained impressive achievements in many fields, it is not always the optimal AI solution. Deep learning models tend to take tremendous volumes of data to feed and build such systems hence when the data is small, deep learning algorithms might not perform properly. Since our dataset is not large, we would also like to experiment with a more traditional approach of machine learning in this subsection.

4.2.1 Data Encoding

In this stage, this research employed the pre-trained Vietnamese SBERT to encode the short news. Vietnamese SBERT is a method that combines sentence encoding technique and the PhoBERT pre-training models described in the previous section, which could transform news data into numeric vectors.

This SBert processing resulted in a list of 1×768 dimensional feature vectors, each of which corresponds to a single news (Fig. 5).

4.2.2 Model Architecture

Support Vector Machine (SVM) [20] and Random Forest (RF) [21] techniques were employed for our machine learning simulation. The SVM classifiers tend to search for the hyperplanes that maximize the margins of a class and use a vector to separate hyperplanes and a bias estimated to minimize the error of the training set. Besides, Random Forest estimator is an ensemble learning method which fits multiple decision tree classifiers/regressors by random selection of features and optimizes by bagging and aggregating the results. We employed Sklearn package to implement our model, using the following hyperparameters:

- For SVM model, we have selected the regularization parameter as 1.0, the degree as 3, the tolerance for stopping criterion as 0.001 and not to enable probability estimates.
- For Random Forest model, we have selected the number of trees in the forest as 100, the maximum depth of the tree as 0, the minimum number of samples required to split an internal node as 2 and the minimum number of samples required to be at a leaf node as 1.

```
array([ 4.31373790e-02,  2.19225794e-01, -4.82123464e-01, -4.21190262e-01,
       -3.04402351e-01,  6.28092766e-01, -4.63526398e-01,  2.00562641e-01,
       -5.14748394e-01, -1.04810923e-01,  4.79191393e-01,  4.99825716e-01,
        2.23641887e-01, -2.89443642e-01, -2.18977422e-01, -1.80746138e-01,
```

Fig. 5 Example of an embedded feature vector

4.3 Data Augmentation

Data augmentation is utilized on the datasets to be tested for RNN models. We employed the following techniques: normal distribution vector addition, random word removal, and random word swapping. The datasets are denoted as follows:

- X_train: dataset used to train the model.
- X_test: dataset used to test the accuracy of the model.
- len(X_train): Number of the data samples used to train the model.
- len(X_test): Number of the data samples used to test the model.

4.3.1 Gaussian Vector Addition

We describe here the steps to augment the data by adding a Gaussian (Normal) distribution vector (denoted "Add method" in Fig. 6):

- Step 1: Divide the initial dataset into the training set and the test sets by a ratio of 9:1.
- Step 2: Add the embedded vector of the training set, which consists of the embedded vectors of all constituent samples, to a random Gaussian vector with size of len(X_train) x 768.
- Step 3: Combine the initial training dataset with that resulted from the adding of a Gaussian random vector.
- Step 4: Apply vector normalization on the combined datasets.

4.3.2 Random Word Removal

Here are the steps to augment data by random word removal (denoted "Remove method" in Fig. 6):

- Step 1: Construct a function for random deletion of n words.
- Step 2: Divide the initial dataset into the training and the test sets with a ratio of 9:1.
- Step 3: Apply the random word deletion function on the training dataset.
- Step 4: Combine the initial training dataset and that resulted from Step 3.

4.3.3 Random Word Swapping

Here are the steps to augment data by random word swapping (denoted "Swap method" in Fig. 6):

- Step 1: Construct a function that swaps a random pair of words.
- Step 2: Divide the initial dataset into the training and the test sets with a ratio of 9:1.

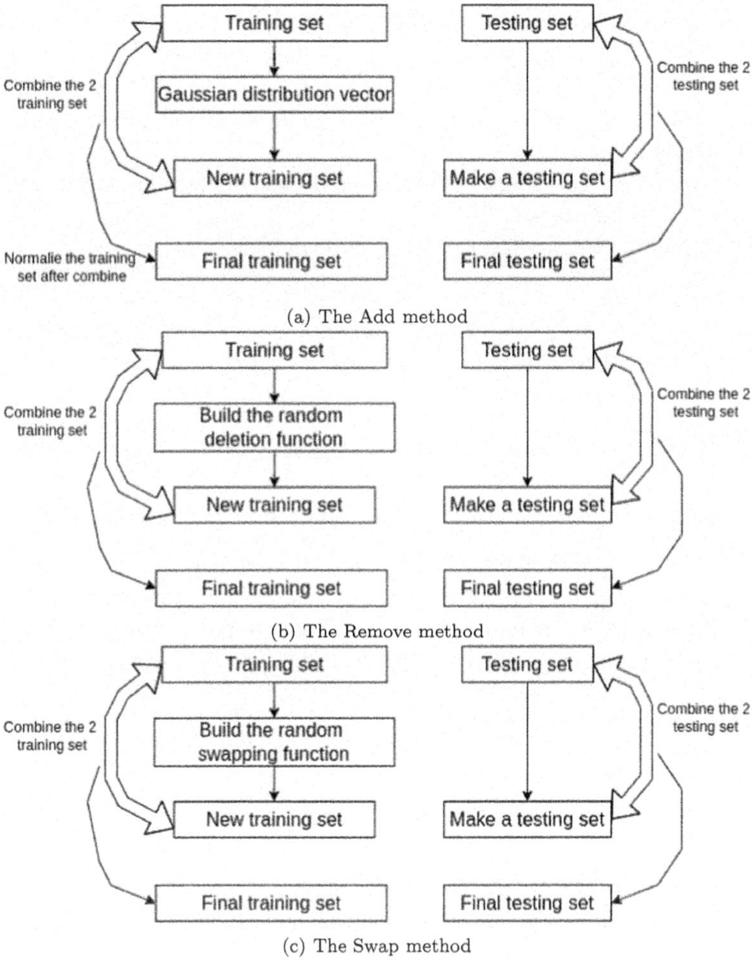

Fig. 6 Data Augmentation

- Step 3: Apply the random word swapping function *n* times on the initial training dataset.
- Step 4: Combine the initial training dataset and that resulted from Step 3.

4.4 Model Ensembling

From the architectures constructed and evaluated in the previous section, we selected the optimal configurations to implement the model ensembling as follows:

- Step 1: Select models with optimal results.
- Step 2: Select the optimal dataset and divide the dataset into 3 parts of a training set, a test set and a validation set.
- Step 3: Train each selected model on the training set and fine tune its parameters on the validation set.
- Step 4: After the validation is complete, apply each finetuned model to predict the trend labels on the test set.
- Step 5: Combine the prediction results of the models.

5 Experimental Results

5.1 Data Splitting

Our data folders are denoted using the following naming conventions:

- 25_7: 25080 samples being labeled by the 7-day average prices.
- 33_7: 33490 samples being labeled by the 7-day average prices.
- 25_3: 25080 samples being labeled by the 3-day average prices.
- 33_3: 33490 samples being labeled by the 3-day average prices.

For the RNN models, datasets were split into training sets, validation set and test sets by the corresponding ratio of 8:1:1. For the machine learning models, training sets and test sets were split by the ratio of 9:1.

5.2 Recurrent Neural Network Results

As shown in both tables in Fig. 7, 7-day dataset models produce more accurate test results than 3-day one. This strongly suggests that the duration of 3 days after posting the news will not be enough for the news to affect the stock price, hence hence longer time would be needed for better analysis. BLSTM_2 and LSTM_2 models in achieve the best training set accuracy of 92.75% and 91.9%, respectively; whereas the LSTM_1 and BLSTM_1 models in achieve the best testing accuracy of 52.27% and 52.03%, respectively.

5.3 Ensembling RNN Results

We ensembled three optimal models, namely RNN_1, LSTM_2, BLSTM_2, which were trained on the 25_7 dataset and obtained the best testing accuracy among all recurrent network models. Figure 8 shows that the ensembling models attained an

	Accuracy (%)					
	Training		Validation		Testing	
	25_3	33_3	25_3	33_3	25_3	33_3
RNN_1	60.62	57.53	49.32	47.57	48.21	48.79
RNN_2	48.91	47.58	45.73	48.13	45.73	48.16
LSTM_1	81.96	79.11	50	47.78	47.72	46.9
LSTM_2	90.19	90.08	47.65	48.7	49.24	48.46
BLSTM_1	81.81	78.58	48.56	49.21	47.48	47.38
BLSTM_2	91.86	91.56	47.01	48.4	47.36	48.64

	Accuracy (%)					
	Training		Validation		Testing	
	25_7	33_7	25_7	33_7	25_7	33_7
RNN_1	63.33	61.28	51.12	50.07	50.03	49.86
RNN_2	49.91	48.46	49.64	48.7	49.64	48.7
LSTM_1	84.1	80.93	50.56	50.43	52.27	50.04
LSTM_2	91.9	89.64	51.54	51.15	51.03	51.5
BLSTM_1	83.14	80.78	49.12	51.63	52.03	50.58
BLSTM_2	92.75	90.74	51.99	51.33	50.08	51.18

Fig. 7 RNN models with 3-day prices (left) and 7-day prices (right)

Fig. 8 Ensemble

Model	Accuracy(%)
RNN_1	50.03%
LSTM_2	51.03%
BLSTM_2	50.08%
Ensemble	52.87%

accuracy of 52.87%, which improves all 3 constituent models RNN_1, LSTM_2, BLSTM_2 by 2.84%, 1.84%, and 2.79%, respectively.

5.4 *Machine Learning Results*

As can be observed, data augmentation is unlikely to improve accuracy much and sometimes gives worse results than not using. Still, the 33490-sample dataset models mostly improve the 25080-sample ones. Without the dataset augmentation, the label augmentation alone improves the accuracy by 2.58 % on average. Random Forest also generally performs better than SVM models. Model using 7-day dataset improves those of 3-day dataset. The best accuracy of 55.4% is obtained using Random Forest on 25080-sample dataset without data augmentation, which clearly improves our previous results of recurrent neural networks (Fig. 9).

Technique augmentation	Accuracy (%)							
	3 days				7 days			
	SVM		Random Forest		SVM		Random Forest	
	25	33	25	33	25	33	25	33
Not using	49.2	53.3	49.4	50.5	52.3	54.9	52.9	55.4
Normal	49.2	53.2	49.5	50.4	48.6	49.5	51.7	48.5
Remove	49.1	53.3	48.4	51.8	51.8	55	51.7	54
Swap	49.2	53.3	48.4	51.4	51.9	54.8	52.3	53.7

Fig. 9 Machine learning model with data augmentation

6 Conclusion

In this work, we predicted stock prices trend from financial news articles using both RNN and machine learning models. We achieved better results with such machine learning models like SVM and Random Forest. Still, we applied ensembling technique for RNN models and obtained improvement of prediction accuracy. To tackle the label data missing due to the lack of stock price data, we proposed an augmentation technique for trend label data, which helps outperform the original dataset on average of 2.58 % of accuracy. There are still different ways to build stock trend prediction models, which we leave as future work. Some of these include combining the sentence encoding with RNN models, experimenting with other distance functions in the label augmentation method, combining various technical indicators with the fundamental analysis approach that might improve the stock prediction in brief windows of time.

References

1. M. Chinas, Efficient market hypothesis: weak form efficiency: an examination of weak form efficiency. Amazon Digital Services LLC-KDP Print US (2018). ISBN: 9789925738366
2. V. Lavrenko et al., Mining of concurrent text and time series, in *KDD-2000 Workshop on Text Mining* (2000)
3. J.D. Thomas, K. Sycara, Integrating genetic algorithms and text learning for financial prediction, in *Data Mining with Evolutionary Algorithms* (2000), pp. 72–75
4. C.-H.L. Lee, A. Liu, W.-S. Chen, Pattern discovery of fuzzy time series for financial prediction. IEEE Trans. Knowl. Data Eng. **18**(5), 613–625 (2006)
5. R.K. Dase, D.D. Pawar, Application of artificial neural network for stock market predictions: a review of literature. Int. J. Mach. Intell. **2**(2), 14–17 (2010)

6. K. Kim, Financial time series forecasting using support vector machines. Neurocomputing **55**(1–2), 307–319 (2003)
7. M.I. Yasef Kaya, M.E. Karsligil, Stock price prediction using financial news articles, in *2nd IEEE International Conference on Information and Financial Engineering* (IEEE, 2010), pp. 478–482
8. A.A. Bhat, S. Sowmya Kamath, Automated stock price prediction and trading framework for nifty intraday trading, in *2013 Fourth International Conference on Computing, Communications and Networking Technologies (ICCCNT)* (IEEE, 2013), pp. 1–6
9. M. Makrehchi, S. Shah, W. Liao, Stock prediction using event-based sentiment analysis, in *IEEE/WIC/ACM International Joint Conferences on Web Intelligence (WI) and Intelligent Agent Technologies (IAT)*, vol. 1 (IEEE, 2013), pp. 337–342
10. B. Narendra et al., Sentiment analysis on movie reviews: a compara tive study of machine learning algorithms and open source technologies. Int. J. Intell. Syst. Appl. 8(8), 66 (2016)
11. R. Schumaker, H. Chen, Textual analysis of stock market prediction using breaking financial news: the AZFin text system. ACM Trans. Inf. Syst. 27 (Feb 2009). https://doi.org/10.1145/1462198.1462204
12. S. Kannan et al., Preprocessing techniques for text mining. Int. J. Comput. Sci. Commun. Netw. 5(1), 7–16 (2014)
13. T. Tran et al., Neu-stock: stock market prediction based on financial news, in *Proceedings of the 2nd International Conference on Human-centered Artificial Intelligence (Computing4Human 2021). CEUR Workshop Proceedings* (Da Nang, Vietnam, 2021)
14. D. Duong, T. Nguyen, M. Dang, Stock market prediction using financial news articles on Ho Chi Minh stock exchange, in *Proceedings of the 10th International Conference on Ubiquitous Information Management and Communication* (2016), pp. 1–6
15. D.Q. Nguyen, A.T. Nguyen, PhoBERT: pre-trained language models for Vietnamese (2020). arXiv:2003.00744
16. T. Mikolov et al., Efficient estimation of word representations in vector space, in *Proceedings of Workshop at ICLR 2013* (2013)
17. A. Sherstinsky, Fundamentals of recurrent neural network (RNN) and long short-term memory (LSTM) network (2018). CoRR arXiv:1808.03314
18. Y. Perwej, The bidirectional long-short-term memory neural network based word retrieval for Arabic documents. Trans. Mach. Learn. Artif. Intell. 3(1), 16–27 (2015). https://doi.org/10.14738/tmlai.31.863. Url: https://hal.archives-ouvertes.fr/hal-03371886
19. K. Zhang et al., Bidirectional long short-term memory for sentiment analysis of Chinese product reviews (2019), pp. 1–4. https://doi.org/10.1109/ICEIEC.2019.8784560
20. M.A. Hearst et al., Support vector machines. IEEE Intell. Syst. Appl. 13(4), 18–28 (1998). https://doi.org/10.1109/5254.708428
21. L. Breiman, Random forests. Mach. Learn. 45(1), 5–32 (2001)

Applications to Transportation Engineering

Lightweight Models' Performances on a Resource-Constrained Device for Traffic Application

Tuan Linh Dang, Duc Loc Le, Trung Hieu Pham, and Xuan Tung Tran

Abstract This paper investigated vehicle and license plate detection tasks in a traffic application using a low-cost embedded hardware platform, NVIDIA Jetson Nano. Different state-of-the-art lightweight models have been customized and examined to provide a comprehensive assessment of implementing deep learning models on Jetson Nano. The vehicle detection task has been tested with CenterNet MobileNetV2 FPN, YOLOv4-tiny, and YOLOv7-tiny, while the license plate detection section investigated WPOD-NET and EfficientNetB0. Experimental results showed that the Jetson Nano-based version had been successfully implemented and obtained similar results as the standard GPU-based version on Google Colab. The vehicle detection was at 70% using the AP@0.5 metric. In addition, YOLOv7-tiny in vehicle detection and WPOD-Net in license plate detection obtained high frames per second (FPS), at 42 FPS and 204 FPS, which could be suitable for real-time application.

Keywords Edge device · Vehicle detection · License plate detection · Lightweight models

T. L. Dang (✉) · D. L. Le · T. H. Pham · X. T. Tran
School of Information and Communications Technology, Hanoi University of Science and Technology, Hanoi 100000, Vietnam
e-mail: linhdt@soict.hust.edu.vn

D. L. Le
e-mail: loc.ld211033M@sis.hust.edu.vn

T. H. Pham
e-mail: hieu.pt194763@sis.hust.edu.vn

X. T. Tran
e-mail: tung.tx194405@sis.hust.edu.vn

N. H. Phuong and V. Kreinovich (eds.), *Deep Learning and Other Soft Computing Techniques*, Studies in Computational Intelligence 1097,
https://doi.org/10.1007/978-3-031-29447-1_16

1 Introduction

Recently, thanks to the development of Artificial Intelligence and traffic surveillance system, many viable solutions have been deployed in practice such as traffic flow detection, anti-breaking law, automated license plate recognition (ALPR), and on-street parking [1]. The intelligent transport system based on cloud computing allows the application to implement complex Deep learning algorithms that need a substantial amount of computing and storage resource [2]. In contrast, high reliance on data transmission and network bandwidth consumption are the main drawbacks of these systems.

From the above-mentioned disadvantages of the cloud computing concept, processing data directly closer to the source, known as edge computing, is increasingly getting interest from the scientific community [3]. To implement machine learning models on edge devices, upgrading hardware computing capabilities and optimizing machine learning algorithms to adapt are required. Several lightweight models had been published [4–6]. The main objectives of these lightweight models remain acceptable accuracy and speed when applied to edge devices.

This paper proposes an approach for license plate detection implemented in a low-cost edge device, Jetson Nano. Our approach includes vehicle detection in the first phase and license plate detection in the second phase. The vehicle detection phase examined three state-of-the-art lightweight models called YOLOv4-tiny, YOLOv7-tiny, and CenterNet MobileNetV2 FPN. In addition, the license plate detection phase investigated WPOD NET and Corner Regression EfficientNetB0 backbone. All models were converted to TensorRT, an NVIDIA GPU accelerator-supported format. To evaluate the performance of deep learning models implemented on Jetson Nano, AP@0.5, AP@0.5:0.95, and FPS were used in the first phase. Mean IoU and FPS were two metrics for measuring performance in the second phase.

This paper is structured as follows. Section 2 describes related works that deal with the recent embedded computer-based traffic monitoring studies, applied computer vision algorithms, and hardware platforms. Proposed methods are presented in Sect. 3. Experiment setup and its results are introduced in Sect. 4. Section 5 concludes our paper.

2 Related work

2.1 Deep Learning-Based Embedded System for Traffic Monitoring Applications

A traffic flow detection algorithm using intelligent video was implemented on NVIDIA Jetson TX2 [7]. To maintain both accuracy and short time delay, a modified YOLOv3 [8] network and DeepSORT were used for vehicle detection and tracking. The final test results indicated that this method achieved an average accuracy of 92.0% and a speed of 37.9 FPS. In addition, a lower-cost NVIDIA embedded computer, Jetson Nano, was used to perform vehicle and pedestrian detection based on several models [9]. It was found that Jetson Nano can handle MobileNetV1, MobileNetV2, and InceptionV2 to produce high accuracy results with fast speed.

An edge-AI-based real-time automated license plate recognition (ER-ALPR) was proposed to implement on two embedded computers Jetson AGX XAVIER [10]. To detect license plate character recognition, YOLOv4-Tiny was used and obtained results with an accuracy rate up to 97% under daytime conditions. In addition, Raspberry Pi and Intel Neural Compute Stick were employed for deploying RPNet, TE2E, and YOLOv3 in ALPR and object detection domains [11]. These deployment concepts had proved capabilities in running machine learning models.

2.2 Algorithms

2.2.1 Vehicle Detection

(a) YOLO Networks

YOLO (You Only Look Once) [12] is a family of models widely used for object detection tasks. Since its publication, many subsequent improved versions of YOLO have been researched and developed by various researchers, with the latest version being YOLOv7 [13]. In this paper, we chose the stripped-down version of YOLOv4 [14] and YOLOv7, called YOLOv4-tiny and YOLOv7-tiny, respectively, to conduct the experiments, given the limited computing resources on edge. Both tiny versions use fewer convolutional layers, YOLO layers, and anchor boxes to make predictions, resulting in a reduction in the number of parameters and FLOPS, allowing for real-time performance on resource-constrained devices.

(b) CenterNet

CenterNet [15], a keypoint-triplets method for object detection published in 2019, is based on CornerNet [16]. The model used three keypoints called the top-left corner, the bottom-right corner, and the center to define a bounding box on the image. Evaluating on MS-COCO [17] dataset, CenterNet [15] achieves 47.0% average precision (AP), which was better than all current one-stage object detection models at the time by at least 4.9%. The paper's authors proposed two modified modules named center pooling to capture recognizable features from regions and cascade corner pooling to acquire both the marginal information and the visual features information of objects.

2.2.2 License Plate Detection

Most existing license plate (LP) detectors can only infer a rectangular region surrounding the LP, but not precisely four corners of the LP itself. This can be acceptable in scenarios where the view of LP is straight and not distorted. But in a real-world traffic setting, images of vehicles captured from surveillance cameras can appear from various angles, making the LP region view oblique. Thus, the content of the LP is not easy to recognize. If only a rectangular bounding box is detected, it would be challenging for later stages (OCR) to process the result. This paper examined two CNN models which can precisely capture four corners of the LP, then used a perspective transformation to "unwarp" the distorted LP into a frontal, rectangular shape.

(a) WPOD-NET
WPOD-NET [18] is a novel CNN architecture aiming to address the task of LP detection in unconstrained scenarios. The network is reported to significantly improve the performance of the whole LP recognition process in settings where the LP's view is oblique and distorted. Given an input image, the output of the network is a tensor with eight channels, corresponding to eight values per pixel position. Two of them are the probabilities of having or not having the LP in that cell. The six remaining values are coefficients to build an affine matrix that transforms a fixed fictional square around that cell into an LP region. After going through Non-max Suppression, the final outputs are rectified using a perspective transformation to "unwarp" the LP region into a frontal view.

(b) EfficientNet
EfficientNets [19] is a series of models developed by a systematically, uniformly compound scaling method. EfficientNet-B0 is the fastest and most lightweight model in the family. This baseline model was obtained by using a neural architecture search algorithm. Using EfficientNet as an extractor, we can solve many different tasks, such as classification or corner regression for license plate detection, which is investigated in this paper.

2.3 Hardware Specification of Edge Embedded System

Our study uses the NVIDIA Jetson Nano Production module as the edge embedded device because of its small size of 69 mm × 45 mm and its reasonable price, as well as excellent software support. The hardware specifications included a 128-core GPU, Quad-core ARM A57 CPU, 4GB of LPDDR4 RAM, and 64GB of storage. It was pre-installed with NVIDIA JetPack SDK, including OS images and optimizing libraries such as CUDA and TensorRT.

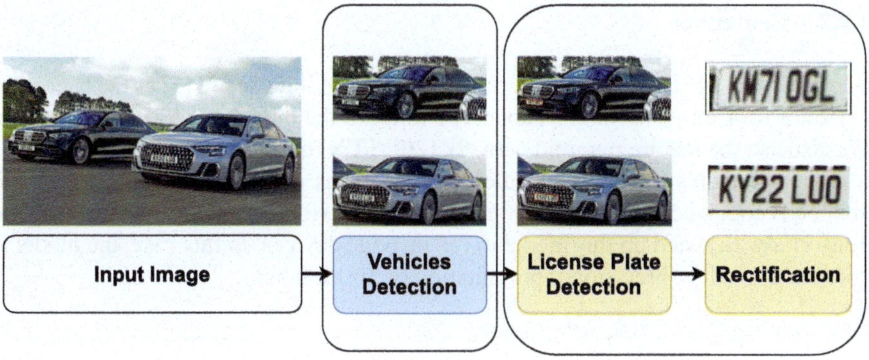

Fig. 1 Our proposed approach

3 Methodology

3.1 Architecture Overview

The overview of our approach is presented in Fig. 1. There are two main components in the system. The first one uses an object detection model to detect vehicles from an input image. Then the second use a deep learning model to detect license plates from regions of interest produced by the earlier stage.

3.2 Vehicle Detection

3.2.1 YOLO Networks

Although YOLOv4-tiny and YOLOv7-tiny had been dramatically compressed, they still could not achieve real-time performance on Jetson Nano when implemented in Darknet or PyTorch, due to limited resources and a non-optimized computing graph for the hardware. Therefore, after training the models on our custom dataset, we converted these two models into TensorRT format to achieve maximum speed on NVIDIA GPUs. YOLOv7-tiny could be easily converted to ONNX format and then to TensorRT, while YOLOv4-tiny has a Mish activation function that is not natively supported. Based on the mathematical definition of Mish function $f(x) = x \times tanh(softplus(x))$, we realized that it could be replaced by Softplus, Tanh, and Mul functions which are supported in TensorRT by default. Parameters of both models are quantized down to 16-bit floating point, which is a good choice for the speed-accuracy trade-off.

3.2.2 CenterNet

Because of its lightweight characteristics, we used MobileNetV2 [5] as the backbone extractor for CenterNet in this paper. The input of the model is a 512×512 image. We also use the feature pyramid network [20] (FPN) design in the extractor. It helps to extract features at different scales, which means objects of different sizes can also be recognized better. Later, we use the TensorRT optimizer integrated inside TensorFlow to speed up this model to run on edge devices. In this case, the model is converted to a low-precision representation of 16-bit floats.

3.3 License Plate Detection

3.3.1 WPOD-NET

As for the WPOD-NET license plate detection model, we kept all the backbone components and modified the network output to better suit our custom dataset. Initially, the model was trained on images containing long, one-line license plates, so the authors only used a single output dimension to unwarp the LP region. In other words, the original model cannot distinguish whether an LP has a square-like aspect ratio or a lengthy, rectangular-like aspect ratio. Thus all square, two-line LPs are resized to the wrong aspect ratio, making them look distorted and unnatural. Although the aspect ratio can be inferred by calculating the width and height of the LP from four output corners, this method is unstable due to the polygon shape of the output and had quite poor performance in our experiments. To address this issue, we modified the original architecture by adding two channels to the network's output, representing the probability of the LP in that image as one-line / two-line. The sigmoid activation function is used for these two channels, and the loss used is Binary Cross Entropy, only taken into account at the cell position where the LP region is presented and is scaled down to 30% before adding to the total loss of the network. We experimented with both approaches to see if modifying the architecture helped with classifying LP type more than just looking at the ratio of edge lengths of the detected polygon.

3.3.2 Corner Regression Model With EfficientNetB0

We use EfficientNetB0 as our backbone extractor in this model. A global average pooling layer is placed at the top of the extractor to convert the extracted features into a vector. Next, we place a 256-dimension fully connected dense layer followed by a final dense layer with eight units. These eight units correspond to the coordinates of four corners of the desired bounding box. This model is also optimized by the TensorRT engine integrated inside the Tensorflow framework to run on edge devices. The number representation of the model is converted to 16-bit floats for faster execution.

4 Experiments

4.1 Datasets

4.1.1 Vehicle Dataset

The vehicle detection dataset used in this paper is combined from the Vietnamese Vehicle dataset (VHD) [21] and the Vehicle Open Image dataset (VOID) [22]. VHD is a dataset of popular Vietnamese vehicles in Ho Chi Minh City at different times. Images, as seen in Fig. 2 are extracted from traffic cameras at various times, backgrounds, and weather conditions, ensuring the diversity of the dataset. All images are labeled with bounding boxes; each box corresponds to one of four classes: motorcycle, car, bus, or truck. There are 12660 images in total, collected from 25 traffic cameras [21]. Figure 3 shows several images from VOID is a dataset of 627 vehicles derived from OpenImages open source computer vision library. This dataset contains vehicles from five classes: motorcycle, car, bus, truck, and ambulance, with 1194 bounding box annotations (1.9 per image by average) and a median image ratio of 1024 $\times$ 571. To merge two datasets, we treat all ambulance images in the second dataset as cars. That means the combined dataset only has four classes: motorcycle, car, bus, and truck [22].

The final dataset includes all the images from the two sources above, split as shown in Table 1.

4.1.2 License Plate Detection Dataset

The license plate detection dataset has a much smaller size compared to the vehicle detection Dataset, with 421 images in total, and is a combination of three sources: 189 images from Cars Dataset [23], 51 images from AOLP Dataset [24], and 181 images from CarTGMT [25] Dataset. Images are samples from three sources to ensure that LPs appear in different backgrounds, views, countries, and conditions. Long, one-line and square, two-line LP are both included. Also, all images in the CarTGMT dataset and 85 images in the Cars Dataset do not have any annotations, so we had to manually label all of them. Examples of the dataset are presented in Fig. 4. Table 2 shows the data splitting in our experiments.

4.2 Results

4.2.1 Vehicles Detection

Results of our three object detection models in the vehicles detection task are shown in Table 3. YOLOv7-tiny gave the best result in both AP and speed with 75.5%

Fig. 2 Images from Vietnamese Vehicle Dataset

AP@0.5 and 42 FPS on Jetson Nano, respectively. On the other hand, CenterNet obtained the worst AP and was the slowest model, but the most lightweight one with only 2.4 million parameters.

4.2.2 License Plate Detection

In detecting license plate task, Table 4 showed our results.

Experimental results demonstrated that WPOD-NET is superior in size, speed, and IoU. With only about 1.6M parameters, the model performed well at around 0.76 mIoU while delivering very high speed on both Google Colab and Jetson Nano. FPS on Jetson Nano is just a bit slower than that on Google Colab, which is impressive given the low-end hardware of the device. The results again demonstrated that TensorRT could give the model a surprising speed boost with little to no accuracy trade-off. Our custom Corner Regression model based on EfficienetB0 also showed comparable performance with slightly lower mIoU but still ensured real-time speed.

As for LP type classification, the results of two approaches: using fixed rules and modifying model architecture, were shown in Table 5. The threshold for the ratio of the longest and second-shortest edge of the detected LP region was set to 1.7. If

Fig. 3 Images from OpenImages's dataset

Table 1 Vehicle detection dataset

	Train	Validation	Test
Vietnamese Vehicle Dataset	12221 images	1183 images	1987 images
OpenImages Vehicle Dataset	439 images	125 images	63 images
Total	12660 images	1308 images	2050 images

that ratio exceeds the chosen threshold, the LP will be classified as a square, two-line. Otherwise, it will be classified as a lengthy, one-line LP. Experiments showed that, with the modified architecture, the model could better recognize the type of LP presented in the image, allowing for better rectification using perspective transform.

Fig. 4 Images from license plate dataset

Table 2 License plate dataset

	Train	Validation	Test
Cars dataset	149 images	20 images	20 images
AOLP dataset	51 images	0 images	0 images
CarTGMT dataset	100 images	40 images	41 images
Total	300 images	60 images	61 images

Table 3 Results of vehicles detection models

Model	Framework	Parameters	Google Colab			Jetson Nano		
			AP@0.5	AP@0.5:0.95	FPS	AP@0.5	AP@0.5:0.95	FPS
YOLOv7-tiny	Pytorch	6.0M	0.747	0.454	143	0.755	0.460	42
YOLOv4-tiny	Darknet	6.1M	0.745	0.396	151	0.716	0.386	25
CenterNet MobileNetV2 FPN	Tensorflow	2.4M	0.695	0.392	81	0.696	0.393	6

Table 4 Results of license plate detection models

Model	Framework	Parameters	Google Colab		Jetson Nano	
			mIoU	FPS	mIoU	FPS
Corner Regression Efficient-NetB0 backbone	Tensorflow	3.5M	0.730	138	0.730	56
WPOD-NET	Tensorflow	1.6M	0.762	260	0.758	204

Table 5 Results of license plate type classification

	Using modified model	Using rule
Correct	55	37
Total	61	61
Accuracy	90.1%	60.6%

5 Conclusions

We researched and tested five deep learning models for traffic-related tasks in this paper. Three of them are for vehicle detection, and two of them are for license plate detection. The YOLOv4-tiny and YOLOv7-tiny models are used directly, while the Centernet-FPN, WPOD-NET, and EfficientnetB0 Corner Regression models are customized to suit the problem better and ensure computation speed.

We implemented and evaluated the performance and speed of all models on our custom dataset for each task. The results showed that, for a mid-range GPU like Tesla T4, the inference speed is very high, and the accuracy is quite good. When deploying on edge devices, specifically Jetson Nano, the speed was reduced quite a lot, but it can still be run in real-time, and the accuracy is almost unchanged, even better in some cases. It could be seen that the Jetson Nano is fully capable of deploying deep learning models and can be effectively utilized for solving many computer vision problems.

A possible direction for future improvement is to test more models and try to customize them to reduce the number of parameters and computation cost. We plan to use the investigated results in this paper to implement a practical application that can solve a real-life problem.

References

1. J. Shashirangana et al., Automated license plate recognition: a survey on methods and techniques. IEEE Access **9**, 11203–11225 (2021). https://doi.org/10.1109/ACCESS.2020.3047929
2. K. Simonyan, A. Zisserman, Very deep convolutional net works for large-scale image recognition (2014). arXiv:1409.1556
3. M. Merenda, C. Porcaro, D. Iero, Edge machine learning for AI-enabled IoT devices: a review. Sensors **20**(9), 2533 (2020)
4. A.G. Howard et al., Mobilenets: efficient convolutional neural net works for mobile vision applications (2017). arXiv:1704.04861
5. M. Sandler et al., Mobilenetv2: inverted residuals and linear bottle necks, in *Proceedings of the IEEE Conference on Computer Vision and Pattern Recognition* (2018), pp. 4510–4520
6. F.N. Iandola et al., SqueezeNet: alexnet-level accuracy with 50x fewer parameters and 5 MB model size (2016). arXiv:1602.07360
7. C. Chen et al., An edge traffic flow detection scheme based on deep learning in an intelligent transportation system. IEEE Trans. Intell. Transp. Syst. **22**(3), 1840–1852 (2020)
8. J. Redmon, A. Farhadi, Yolov3: an incremental improvement (2018). arXiv:1804.02767
9. L. Barba-Guaman, J. Eugenio Naranjo, A. Ortiz, Deep learning framework for vehicle and pedestrian detection in rural roads on an embedded GPU. Electronics **9**(4), 589 (2020)
10. C.-J. Lin, C.-C. Chuang, H.-Y. Lin, Edge-AI-based real-time automated license plate recognition system. Appl. Sci. **12**(3), 1445 (2022)
11. J. Shashirangana et al., License plate recognition using neural architecture search for edge devices. Int. J. Intell. Syst. (2021). https://doi.org/10.1002/int.22471
12. J. Redmon et al., You only look once: unified, real-time object detection, in *Proceedings of the IEEE Conference on Computer Vision and Pattern Recognition* (2016), pp. 779–788
13. C.-Y. Wang, A. Bochkovskiy, H.-Y. Mark Liao, YOLOv7: trainable bag-of-freebies sets new state-of-the-art for real-time object detectors (2022). arXiv:2207.02696
14. A. Bochkovskiy, C.-Y. Wang, H.-Y. Mark Liao, Yolov4: optimal speed and accuracy of object detection (2020). arXiv:2004.10934
15. K. Duan et al., Centernet: keypoint triplets for object detection, in *Proceedings of the IEEE/CVF International Conference on Computer Vision* (2019), pp. 6569–6578
16. H. Law, J. Deng, Cornernet: detecting objects as paired keypoints, in *Proceedings of the European Conference on Computer Vision (ECCV)* (2018), pp. 734–750
17. T.-Y. Lin et al., Microsoft COCO: common objects in context, in *European Conference on Computer Vision* (Springer, 2014), pp. 740–755
18. S.M. Silva, C.R. Jung, License plate detection and recognition in unconstrained scenarios, in *Proceedings of the European Conference on Computer Vision (ECCV)* (2018), pp. 580–596
19. M. Tan, Q. Le, Efficientnet: rethinking model scaling for convolutional neural networks, in *Proceedings of International Conference on Machine Learning. PMLR* (2019), pp 6105–6114
20. T.-Y. Lin et al., Feature pyramid networks for object detection, in *Proceedings of the IEEE Conference on Computer Vision and Pattern Recognition* (2017), pp. 2117–2125
21. Vietnamese Vehicles Dataset. https://www.kaggle.com/datasets/duongtran1909/vietnamese-vehicles-dataset. Accessed 06 Aug 2022
22. Vehicles-OpenImages Dataset. https://public.roboflow.com/object-detection/vehicles-openimages. Accessed 07 Aug 2022
23. Cars Dataset, Standford AI Hub. http://ai.stanford.edu/~jkrause/cars/car_dataset.html. Accessed 07 Aug 2022
24. Application-Oriented License Plate Recognition Dataset. https://github.com/AvLab-CV/AOLP. Accessed 07 Aug 2022
25. CarTGMT Dataset. https://drive.google.com/file/d/1U5ebTzW2c_sVVTCSX1QH-ZJFpLijMdUv/view. Accessed 07 Aug 2022

IT2-Neuro-Fuzzy Wavelet Network with Jordan Feedback Structure for the Control of Aerial Robotic Vehicles with External Disturbances

Rahul Kumar, Uday Pratap Singh, Arun Bali, and Siddharth Singh Chouhan

Abstract Taking into account the various challenges due to uncertainties and disturbances while operating with aerial robotic vehicles (ARVs) in a harsh and extreme environment, a novel adaptive control scheme named Interval-Type-2 Neuro-Fuzzy Wavelet (IT2-NFW) network with Jordan feedback structure has been proposed in this work. The ARVs considered in this study are under the influence of external disturbances due to which the ARV system becomes uncertain and hence becomes very difficult to control. The suggested control strategy can manage system uncertainty more efficiently. With high computational capacity, little computational load, and a quick convergence rate, the proposed IT2-NFW control method can accurately simulate system uncertainties and track the reference trajectory. The controller's stability has been demonstrated by means of Lyapunov's method, and the controller's assured convergence has been demonstrated. Finally, the controller's effectiveness and efficiency have been demonstrated by controlling an aerial robotic vehicle.

1 Introduction

Aerial robotic vehicles (ARVs) have drawn great attention from researchers because of their popularity and incredible applications in military and other civilian tasks like reconnaissance and surveillance [1], search and rescue operations [2], agricultural tasks [3], goods delivery tasks [4], etc. In addition to these applications, the applicability of ARVs in various aspects of life is growing widely. With this increasing applicability of ARVs, there are also arising various challenges in dealing with them. Since the operating domain of an ARV is full of uncertainties due to external dis-

R. Kumar · U. P. Singh (✉) · A. Bali
School of Mathematics, Shri Mata Vaishno Devi University,
Katra 182320, Jammu and Kashmir, India
e-mail: usinghiitg@gmail.com

S. S. Chouhan
School of Computing Science and Engineering, VIT Bhopal University, Bhopal 466114, Madhya Pradesh, India

N. H. Phuong and V. Kreinovich (eds.), *Deep Learning and Other Soft Computing Techniques*, Studies in Computational Intelligence 1097,
https://doi.org/10.1007/978-3-031-29447-1_17

turbances [5], the control of an ARV in such a situation becomes a challenge. Thus carrying out various operations with ARVs in an uncertain environment requires a very efficient control scheme that is capable of dealing with uncertainties effectively; otherwise, the operation may result in failure. Although there exist some classical PID controllers [6–8] for controlling these vehicles but they are not capable to handle uncertain environmental conditions and disturbances. Being linear controllers, these cannot handle nonlinear characteristics due to the nonlinear and changing dynamics of the ARVs and hence lacks measurement accuracy. To cope with this challenge, numerous modern control schemes using neural networks (NNs) and fuzzy logic have also been proposed [9–11].

For instance, [12] presented an ANN-based Proportional derivative approach for trajectory tracking and secure landing of UAVs and the proposed approach is capable of learning unmodelled system dynamics. In [13], a backstepping-based adaptive control scheme has been presented for a UAV equipped with a robotic manipulator. In this approach, the overall system is reduced to a simplified lower dimensional model and focussed on controlling UAV independently of robotic manipulator while [14] focussed on UAVs equipped with multi-link manipulators and proposed a novel scheme by proposing linearizing control laws for nonlinear feedback loops. In addition to these, [15, 16] focused on UAVs with associated payloads where [15] investigated the problem of leaving payloads by UAVs to specified locations using the model reference approach of adaptive control while [16] studied the problem of under actuations in UAVs due to associated payloads by using energy function based control.

Although these results are fruitful, the problem with these control schemes is that they are not capable of accurately modeling and measuring system uncertainties. Also, since the external disturbances are uncertain and can occur at any time in any form without a priori knowledge, hence cannot be accurately expressed as closed-form mathematical expressions. Therefore, it is more convenient to express them in the form of Gaussian white noise.

Motivated by this extensive discussion and to resolve this uncertainty problem, a novel control scheme inspired by the one presented in [17] but differs from [17] in the sense of feedback structure and is named interval-type-2 neuro-fuzzy wavelet (IT2-NFW) network controller utilizing Jordan feedback structure has been proposed in this work. Inspired by work in [18–20], a modern swarm intelligence-based optimization technique called chaos cat swarm optimization is used to initialize the controller's initial weights and other parameters for better performance and quick convergence. This technique outperforms better in the case of multi-objective optimization conventional approaches such as the grey wolf, gravitational search, and particle swarm optimization [21]. The suggested controller has two parts: an antecedent component that uses Jordan Wavelet (JW) NN and a consequent part that uses interval-type-2 fuzzy membership functions. Jordan neural networks (JNNs) have the ability to manage external time delays in feedback, so the JWNN with self-loops gives the controller a special feedback structure and is effective at handling feedback delays, while the IT2-fuzzy membership function allows it to model system uncertainties accurately with little computational work and quick convergence, with self-loops acting as memory units. Therefore, the suggested controller has a large computing capacity and a low computational burden, allowing it to manage quick alterations in the process variables and configuration. Additionally, it has the capacity to recall system's past information.

As a quick recap of this work's major contributions, we have:

- The novelty of this work is the proposed interval-type-2 neuro-fuzzy wavelet (IT2-NFW) controller which can effectively deal with uncertainties in the system.
- The suggested control strategy has the capacity to retain the system's past information to handle quick alterations in the configuration and variables/parameters of the process to be controlled with high level of computing efficiency and minimal amount of computing load.
- The suggested control strategy has been utilized to control the nonlinear dynamics of ARVs in the presence of external disturbances and in comparison to previous control techniques, improved outcomes are attained.

The remaining portion of this article's work is explained in the following manner. In the second section, Sects. 2.1 and 2.2 present the necessary preliminary information, and the controller's stability respectively with the stability analysis in Sect. 2.3. The implementation of the work is given in Sect. 3 with ARV description and simulation in Sects. 3.1 and 3.2 respectively and finally, Sect. 4 provides the work's conclusion.

2 Interval-Type-2 Neuro-Fuzzy Wavelet control

2.1 Preliminaries

1. **Fuzzy Neuron**: A *fuzzy neuron* is an artificial neuron designed by introducing fuzzy concepts at different levels of mathematical operations in a simple artificial neuron [22]. In this study, in order to accurately model system uncertainties, a fuzzy neuron has been designed as follows and can be seen in Fig. 1. The proposed fuzzy neuron consists of n inputs y_is with weights w_is for $i = 1$ to n and type-2 fuzzy membership function (MF) in place of conventional activation function. The net input y to this neuron is collected at the summing junction which is then passed to the membership node consisting of type-2 MF which gives the output A, where A is a fuzzy term associated with the net input y and each input-output value can be associated with the membership value of a fuzzy concept.
2. **Type-2 membership function**: A type-2 fuzzy membership function μ_{A_i} is the one in which membership values of elements of a fuzzy set A is itself a fuzzy set, i.e, membership values are further given by a membership function [23].
3. **Aerial robotic Vehicle**: A new type of extremely intelligent small flying robotic device that can be fully operated remotely and without the use of a human pilot is known as an aerial robotic vehicle. It comprises an electronic hub with arms holding propellers for lift and a power source put on it, allowing it to fly [24, 25].

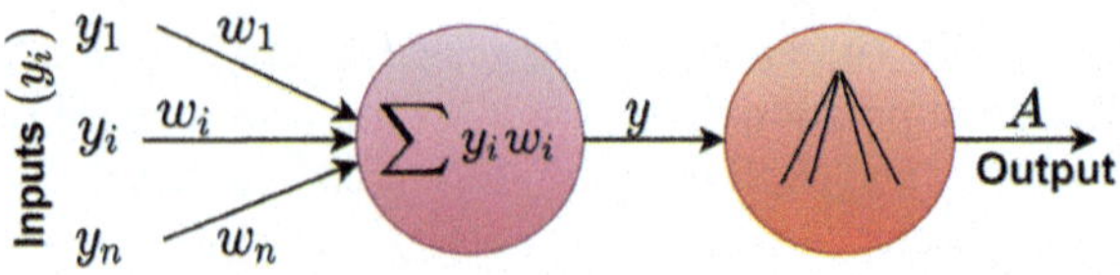

Fig. 1 Fuzzy neuron

4. **Chaos Cat Swarm Optimization**: A new chaos cat swarm optimization (CCSO) algorithm suggested by [21] is used to enhance the convergence of the proposed control scheme. CCSO is an enhanced version of Cat swarm optimization (CSO). The CSO has issues with immaturity and poor tracking accuracy, making it vulnerable to getting trapped in local extrema and CCSO is free from all these problems. Despite the existence of numerous additional iconic algorithms, CSO is a superior method for resolving multi-objective issues. Hence, in this study, CCSO is utilized to optimise the initial parameters of the IT2-NFW controller due to its multi-objective approach. More details on CCSO can be found in [21].

2.2 *Controller Design*

The proposed controller named Interval-Type-2 Neuro-Fuzzy Wavelet (IT2-NFW) Network controller utilizing Jordan feedback structure is a novel hybrid-computing-based neuro-fuzzy controller with feedback structure inspired from the structure of JNN and consists of five layers as shown in the Fig. 2.

The structure of the controller is as:

Layer 1: This layer is the input layer consisting of a single node with linear activation function. According to the architecture of the controller, the output of layer 1 routed to layer 2 and 4.

Layer 2: In this layer, the fuzzy rules are defined and consists of fuzzy neurons utilizing type-2 MFs, the purpose of which is to accurately model system uncertainties. This layer's node count corresponds to the total number of rules which collectively forms the rule-base and are stated in the following way
$R_n : IF\ yis\ A_1,\ OR\ y\ is\ A_2, \cdots,\ yisA_i\ THEN\ v_i\ is\ U_i.$
Here y denotes the input which is given to the IT2-NFW controller and A_i, for $i = 1 : n$ is the lingual term characterized by the fuzzy MF μ_{A_i}. The output of this layer is given to layer 3 which results in the conversion of type-2 MFs into IT-2 MFs and hence makes the overall network capable of efficiently dealing with uncertainties associated with real-life engineering systems with manageable computational load, in contrast to simple type-1 and type-2 MFs.

Layer 3: This layer defines the corresponding lower as well as upper degree of membership for the system input, as shown below in (1)

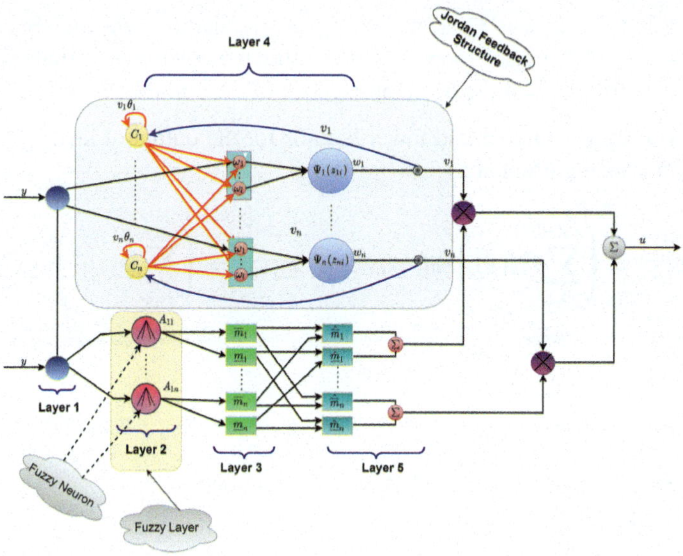

Fig. 2 IT2-NFEW network

$$\overline{m}_{A_i}(y) = exp\left(\frac{-(y-c_i)^2}{\overline{\sigma}_i^2}\right) \text{ and } \underline{m}_{A_i}(y) = exp\left(\frac{-(y-c_i)^2}{\underline{\sigma}_i^2}\right), \quad i = 1:n. \tag{1}$$

Layer 4: In this layer, a Jordan Wavelet NN is designed by integrating together two sub-layers namely the context layer and the wavelet layer. The wavelet layer is made up of n blocks, each of which has a count l number of wavelet neurons. The context layer is composed of nodes $C_1, C_2, \cdots, C_n$ that plays the role of extremely effective memory units for the network. Here, the following wavelet activation function is employed that is defined as

$$\omega(z_{ij}) = z_{ij} * exp\left(-\frac{1}{2}z_{ij}^2\right), i = 1:n, j = 1:l \tag{2}$$

where $z_{ij} = \frac{\xi_{ij}-t_{ij}}{d_{ij}}$, with t_{ij} and d_{ij} being wavelet parameters defining shift and scaling respectively.

The wavelet layer's input that is denoted by ξ_{ij} is stated as

$$\xi_{ij} = y + \sum_{i=1}^{n} C_i(k), \tag{3}$$

where $C_i(k)$ denotes the output of the ith node of the consequent layer at time-step k such that

$$C_i(k) = v_i(k-1) + \theta_i C_i(k-1), \tag{4}$$

where $v_i(k-1) = \Psi_i(k-1)w_i$, $\Psi_i = \prod_{i=1}^{n} \omega(z_{ij})$ and θ_i is the weight of the context layer's ith self-feedback loop. All these initial parameters are initialised by using the CCSO algorithm for the quick convergence of the controller.

Layer 5: The value of upper and lower bounds for the output of layer 3 are defined in this layer which are stated as

$$\hat{\overline{m}}_i(y) = \left(\sum_{j=1}^{n} \overline{m}_k(y)\right) \overline{m}_i(y) \quad \text{and} \quad \hat{\underline{m}}_i(y) = \left(\sum_{j=1}^{n} \underline{m}_k(y)\right) \underline{m}_i(y) \tag{5}$$

where

$$\overline{m}_i(y) = \overline{m}_{A_1} * \overline{m}_{A_2} * \cdots * \overline{m}_{A_n} \quad \text{and} \quad \underline{m}_i(y) = \underline{m}_{A_1} * \underline{m}_{A_2} * \cdots * \underline{m}_{A_n}$$

and the symbol $*$ stands for the AND (min) operation.

Now the IT2-NFW controller's output is given as

$$u = \sum_{i=1}^{n} v_i \left(c\hat{\underline{m}}_i(y) + (1-c)\hat{\overline{m}}_i(y)\right) \tag{6}$$

where each rule's contribution to the output is indicated by the letter c.

2.3 *Stabilty Analysis*

Firstly, in order to compute the tracking error, a control cost function is defined as follows to demonstrate the stability of the suggested controller.

$$\epsilon_e(k) = \frac{1}{2}\left[\epsilon_{e,1}^2(k) + \epsilon_{e,2}^2(k)\right] \tag{7}$$

where $\epsilon_{e,1}^2(k) = (y_d(k) - y(k))^2$ and $\epsilon_{e,2}^2(k) = Au^2(k)$ with A being penalty factor. Also, define the control law as

$$u = \sum_{i=1}^{n} v_i \left(c\hat{\underline{m}}_i(y) + (1-c)\hat{\overline{m}}_i(y)\right) \tag{8}$$

and adaptive law as

$$s^r(k+1) = s^r(k) + \gamma^r \left(\frac{\partial \epsilon_e(k)}{\partial s^l(k)}\right), \quad r = 1:4 \tag{9}$$

where $\gamma^r = [\gamma^w, \gamma^\theta, \gamma^d, \gamma^t]$ is the learning rate corressponding to design parameters $s^r = [s^1, s^2, s^3, s^4] = [w_i, \theta_i, d_{ij}, t_{ij}]$ for $i = 1 : n$, $j = 1 : l$ of the controller. The following theorem is introduced to prove the stability of the proposed controller.

Theorem 1 *The convergence of the proposed IT2-NFW controller is guaranteed for the control law defined as in (8), adaptive law in (9) and control cost function (7) with the maximum learning rate* $\frac{1}{\Gamma}$, *where* $\Gamma = (\alpha^2 + \beta^2)\left\{max \left\| \frac{\partial u(k)}{\partial s^r(k)} \right\| \right\}^2 > 0$ *[26].*

2.4 Complexity Analysis

For the easy understanding of the overall control process following algorithm is presented.

Algorithm 1 Pseudo-code for entire control process

1: Start: $k = 0$.
2: Initialize the states of the UAV system (10) and set y_d.
3: Initiliaze weights & wavelet parameters of the IT2-NFW network using CCSO.
4: Calculate the output y for UAV (10) and error according to (7).
5: **for** $k = 0 : t$ **do**
6:    Calculate u according to (8).
7:    Calculate the output y for UAV (10).
8:    Calculate error according to (7)
9:    **if** Error < threshold **then**
10:       **return** y
11:       break
12:    **else**
13:       Adjust network parameters according to (9).
14:    **end if**
15:    Go back to step 6.
16: **end for**

In view of the entire control process, the computational complexity of the close-loop control depends on the three factors: (a) complexity of CCSO, (b) complexity of IT2-NFW network and (c) the number of repetitions of the loop to compute the error function (7).

For a population of m cats, the total number of computations required to initialize the j number of parameters of IT2-NFW network is given by $O(TmM)$, where T is maximum number of iterations and $d = 1 : M$ with d being dimension of population space. In addition to this, the computational complexity of the IT2-NFW network depends on the complexity of antecedent and consequent part. According to the structure of the antecedent part, the input layer requires $O(n)$ computations while layer 2, 3 and 5 requires $O(2n)$, $O(n^2)$ computational steps respectively. Therefore, the total number of computations in antecedent part are given by $O(n^2)$, where n is the number of nodes in the layer 2. On the other hand, the consequent part of IT2-NFW network requires $O(n^2l)$ computational steps, where l is the number of nodes in each block of wavelet layer. There the total complexity of IT2-NFW network is given as $O(n^2) + O(n^2l) = O(n^2l)$. If t is the maximum number of iterations to compute the error function (7) and control signal (8), then the complexity of IT2-NFW network becomes $O(tn^2l)$.

If $l << n$, then it becomes $O(tn^2)$.

Hence, the total computational complexity of the entire closed-loop controller is given as $O(TmM) + O(tn^2)$.

3 Control of Aerial Robotic Vehicle (ARV)

In order to demonstrate the effectiveness and real-time applicability of the suggested control strategy, the scheme is implemented on an ARV for its control. The control signal for the ARV is provided by the output of the IT2-NFW controller, and the control mechanism for the ARV using the IT2-NFW scheme is shown in Fig. 3.

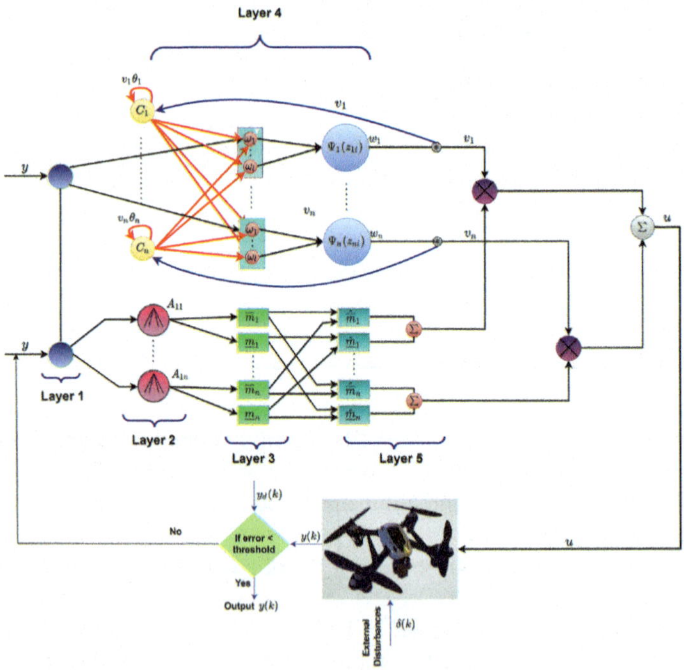

Fig. 3 Control of ARV using IT2-NFW controller

3.1 ARV System Description

The nonlinear dynamical state-space model of a quadrotor aerial robotic vehicle is described as

$$
\begin{aligned}
\chi_1(k+1) &= \chi_1(k) + \eta_1[\chi_2(k) + \delta_1(k)] \\
\chi_2(k+1) &= \chi_2(k) + \eta_2[\frac{I_{yy} - I_{zz}}{I_{xx}}\chi_4(k)\chi_6(k) - I_r\Omega_r\frac{\chi_4(k)}{I_{xx}} \\
&\quad + \frac{\tau_1}{I_{xx}} + \delta_2(k)] \\
\chi_3(k+1) &= \chi_3(k) + \eta_3[\chi_4(k) + \delta_3(k)] \\
\chi_4(k+1) &= \chi_4(k) + \eta_4[\frac{I_{zz} - I_{xx}}{I_{yy}}\chi_2(k)\chi_6(k) + I_r\Omega_r\frac{\chi_2(k)}{I_{yy}} \\
&\quad + \frac{\tau_2}{I_{yy}} + \delta_4(k)] \\
\chi_5(k+1) &= \chi_5(k) + \eta_5[\chi_6(k) + \delta_5(k)] \\
\chi_6(k+1) &= \chi_6(k) + \eta_6[\frac{I_{xx} - I_{yy}}{I_{zz}}\chi_2(k)\chi_4(k) + \frac{\tau_3}{I_{zz}} + \delta_6(k)]
\end{aligned} \tag{10}
$$

where roll (rotation about the x-axis), pitch (rotation about the y-axis), and yaw (rotation about the z-axis) are denoted by the state variables χ_1, χ_3 and χ_5 respectively while the variation in these respective parameters is denoted by the state variables χ_2, χ_4 and χ_6. Propeller inertia and relative speed are shown, respectively, by the symbols I_r and Ω_r. With τ_1, τ_2 and τ_3 representing moments of rotation in all the three axes (x,y & z-axis) respectively, I_{xx}, I_{yy} and I_{zz} are the respective parameters denoting inertia values in all the three directions as per the structural frame of the ARV respectively. Here, in this simulation, τ_1 and τ_2 are kept constant while τ_3 is treated as the input to the ARV, i.e $\tau_3 = u(v(k))$. External disturbances in the roll, pitch, and yaw process cause the ARV to be unreliable and uncertain. These disturbances are denoted by $\delta_i, i = 1:6$ and are taken in the form of gaussian white noise satisfying statistical properties $E(\delta_i) = 0$ and $E(\delta_i^2) = 1$ and is shown in Fig. 5. The working of overall closed loop system can be well understood through a simple flow chart shown in Fig. 4.

3.2 Simulation

The parameters of the ARV system for the simulation are choosen as $I_{xx} = 7.72 \times 10^{-2}$, $I_{yy} = 7.64 \times 10^{-2}$, $I_{zz} = 0.1031$, $I_r = 1.8 \times 10^{-5}$, $\Omega_r = 1$, $\tau_1 = 0.5$ and $\tau_2 = 0.4$ with $\eta_i = 1$, $\forall i = 1:6$.

The initial states of the ARV are selected as $[\chi_1, \chi_2, \chi_3, \chi_4, \chi_5, \chi_6] = [0, 0, 0, 0, 0, 0]$ and the overall ARV system is controlled by utilizing the proposed IT2-NFW

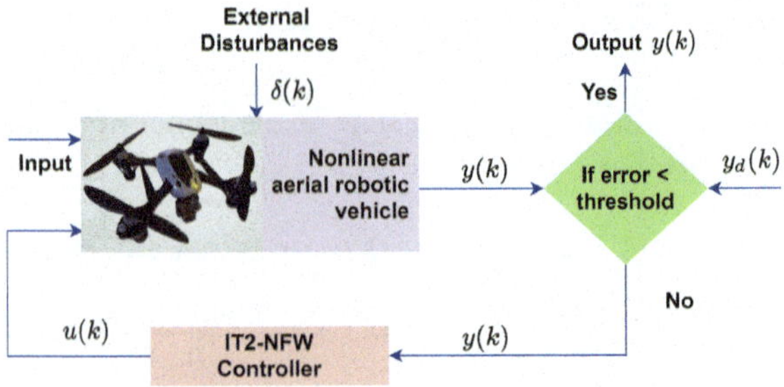

Fig. 4 Overall control process

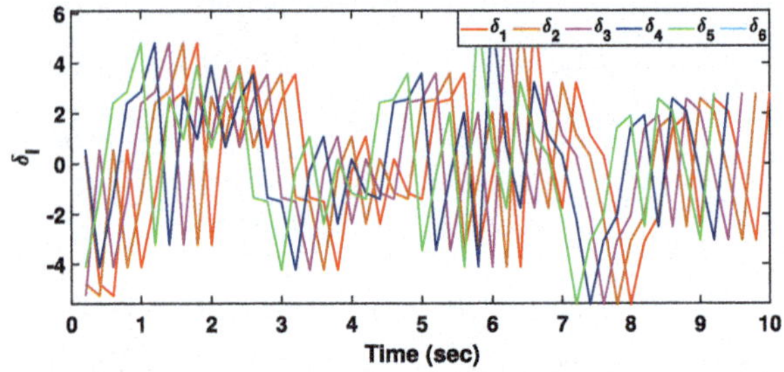

Fig. 5 External disturbance-gaussian white noise

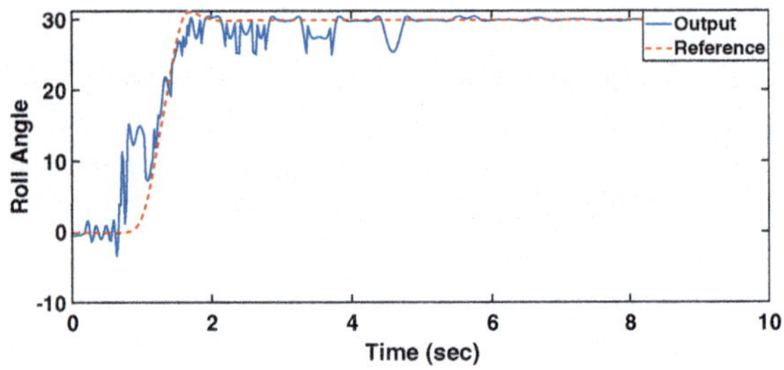

Fig. 6 Tracking of desired trajectory for roll angle

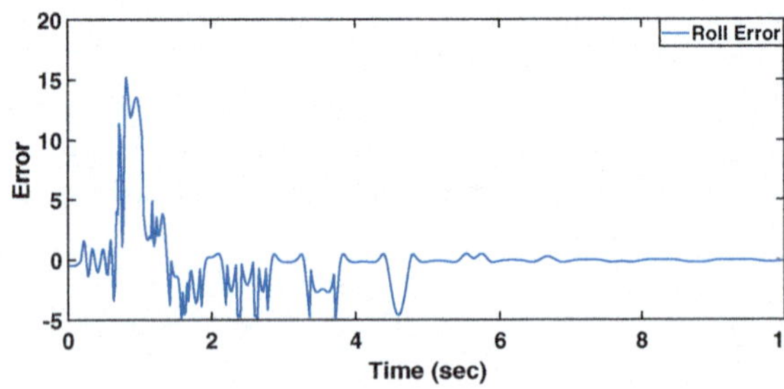

Fig. 7 Error in tracking desired trajectory for roll angle

control scheme. The performance of the so far proposed control scheme can be seen and analyzed from the Figs. 6, 7, 8, 9, 10, 11. Figures 6, 8, and 10 demonstrate the proposed IT2-NFW controller's tracking performance for the roll, pitch, and yaw angles of the ARV respectively while Figs. 7, 9, and 11 indicate the corresponding tracking error.

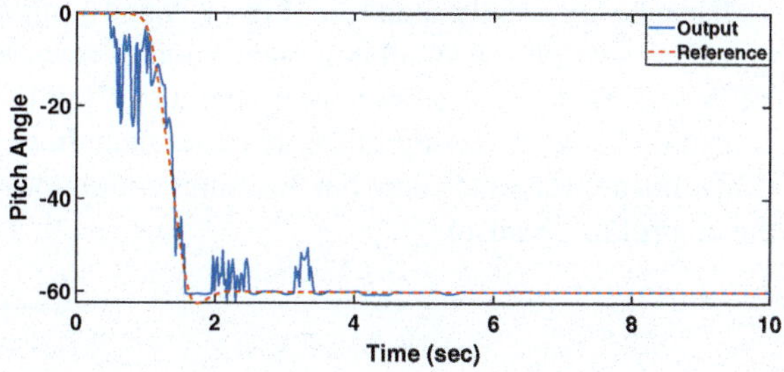

Fig. 8 Tracking of desired trajectory for pitch angle

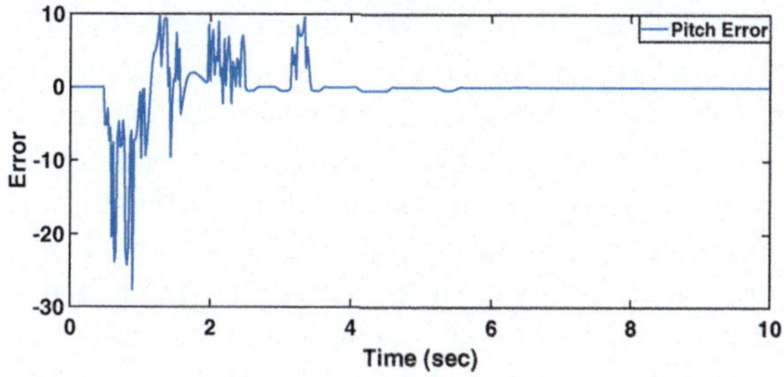

Fig. 9 Error in tracking desired trajectory for roll angle

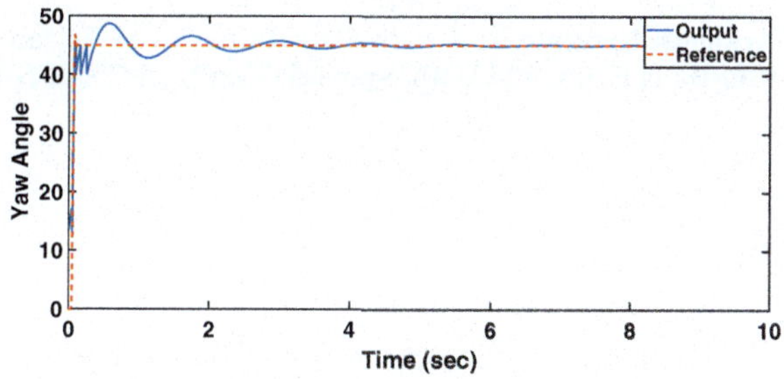

Fig. 10 Tracking of desired trajectory for yaw angle

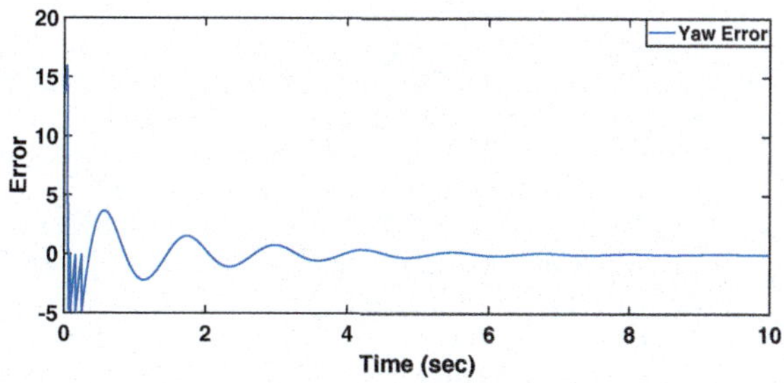

Fig. 11 Error in tracking desired trajectory for roll angle

It can be apparently seen from Figs. 7, 9 and 11 that the controller's tracking error quickly approaches a small neighbourhood of zero. When the performance of IT2-NFW is compared to the work presented in [27], it is found that the results obtained by using IT2-NFW are significantly better than those in [27] as the tracking error in [27] is not convergent to zero but fluctuates continuously in fixed intervals throughout the control mechanism.

4 Conclusion

In this work, a novel adaptive neuro-fuzzy network control scheme has been proposed for the control of aerial robotic vehicles under external disturbances. Due to the presence of external disturbances, the overall system becomes more complex and uncertain to be manageable. The proposed control strategy's efficiency in handling uncertainties with low computing load and rapid convergence enables the ARVs to operate efficiently even in extreme environmental conditions. The convergence of the controller is proved using Lyapunov theory with error converging in a small neighborhood of zero and can be seen in the simulation section.

Acknowledgements This work has been funded by Science and Engineering Research Board (SERB), India under MATRICS project Grant No. MTR/2021/000478.

References

1. T. Kopfstedt, M. Mukai, M. Fujita, C. Ament, Control of formations of uavs for surveillance and reconnaissance missions. IFAC Proc. Vol. **41**(2), 5161–5166 (2008)
2. M. Silvagni, A. Tonoli, E. Zenerino, M. Chiaberge, Multipurpose uav for search and rescue operations in mountain avalanche events. Geomat. Nat. Haz. Risk **8**(1), 18–33 (2017)
3. J. del Cerro, C. Cruz Ulloa, A. Barrientos, J. de León Rivas, Unmanned aerial vehicles in agriculture: a survey. Agronomy 11(2), 203 (2021)
4. B.D. Song, K. Park, J. Kim, Persistent uav delivery logistics: Milp formulation and efficient heuristic. Comput. Ind. Eng. **120**, 418–428 (2018)
5. H.A. Foudeh, P. Luk, J. Whidborne, Application of norm optimal iterative learning control to quadrotor unmanned aerial vehicle for monitoring overhead power system. Energies **13**(12), 3223 (2020)
6. P.E. Pounds, D.R. Bersak, A.M. Dollar, Stability of small-scale uav helicopters and quadrotors with added payload mass under pid control. Auton. Robot. **33**(1), 129–142 (2012)
7. I. Sadeghzadeh, A. Mehta, A. Chamseddine, Y. Zhang, Active fault tolerant control of a quadrotor uav based on gainscheduled pid control, in *2012 25th IEEE Canadian conference on electrical and computer engineering (CCECE)* (IEEE, 2012), pp. 1–4
8. P. Shao, Z. Zhou, S. Ma, L. Bin, Structural robust gain-scheduled pid control and application on a morphing wing uav, in *2017 36th Chinese Control Conference (CCC)* (IEEE, 2017), pp. 3236–3241

9. Q. Lin, Z. Cai, Y. Wang, J. Yang, L. Chen, Adaptive flight control design for quadrotor uav based on dynamic inversion and neural networks, in *2013 Third International Conference on Instrumentation, Measurement, Computer, Communication and Control* (IEEE, 2013), pp. 1461–1466
10. B. Zhao, B. Xian, Y. Zhang, X. Zhang, Nonlinear robust adaptive tracking control of a quadrotor uav via immersion and invariance methodology. IEEE Trans. Ind. Electron. **62**(5), 2891–2902 (2014)
11. L. Xuehui, L. Jingtao, Research on adaptive attitude control of quadrotor uav, in *2017 29th Chinese Control And Decision Conference (CCDC)* (IEEE, 2017), pp. 5904–5909
12. S. Patel, A. Sarabakha, D. Kircali, E. Kayacan, An intelligent hybrid artificial neural network-based approach for control of aerial robots. J. Intell. Robot. Syst. **97**(2), 387–398 (2020)
13. D. Chaikalis, F. Khorrami, A. Tzes, Adaptive control approaches for an unmanned aerial manipulation system, in *2020 International Conference on Unmanned Aircraft Systems (ICUAS)* (IEEE, 2020), pp. 498–503
14. H. Beikzadeh, G. Liu, Trajectory tracking of quadrotor flying manipulators using l1 adaptive control. J. Franklin Inst. **355**(14), 6239–6261 (2018)
15. A. Altan, Ö. Aslan, R. Hacıoğlu, Model reference adaptive control of load transporting system on unmanned aerial vehicle, in *2018 6th International Conference on Control Engineering & Information Technology (CEIT)* (IEEE, 2018), pp. 1–5
16. S. Yang, B. Xian, Energy-based nonlinear adaptive control design for the quadrotor uav system with a suspended payload. IEEE Trans. Ind. Electron. **67**(3), 2054–2064 (2019)
17. Z. Sheikhlar, M. Hedayati, A.D. Tafti, H.F. Farahani, Fuzzy elman wavelet network: applications to function approximation, system identification, and power system control. Inf. Sci. **583**, 306–331 (2022)
18. R. Kumar, U.P. Singh, A. Bali, K. Raj, Hybrid neural network control for uncertain nonlinear discrete-time systems with bounded disturbance. Wirel. Pers. Commun. 1–20 (2022)
19. A. Bali, U. Pratap Singh, R. Kumar, K. Raj, Hybrid neural network control for nonlinear continuous time systems with time delays and dead zone input. Int. J. Adapt. Control Signal Process. **36**(6), 1439–1459 (2022)
20. R. Kumar, U.P. Singh, A. Bali, K. Raj, Hybrid neural network controller for uncertain nonlinear discrete-time systems with non-symmetric dead zone and unknown disturbances. Int. J. Control 1–9 (2022)
21. J. Wang, H. Yu, A new chaos cat swarm optimization algorithm based on saliency gradient for power equipment infrared and visible images registration. Evol. Intell. 1–15 (2022)
22. H.T. Nguyen, C. Walker, E.A. Walker, *A First Course in Fuzzy Logic* (Chapman and Hall/CRC, 2018)
23. J.M. Mendel, R.B. John, Type-2 fuzzy sets made simple. IEEE Trans. Fuzzy Syst. **10**(2), 117–127 (2002)
24. E. Feron, E.N. Johnson, *Aerial Robotics* (Springer, Berlin. Heidelberg, 2008), pp. 1009–1029
25. D. Hodgkinson, R. Johnston, *Aviation Law and Drones: Unmanned Aircraft and the Future of Aviation* (Routledge, 2018)
26. R. Kumar, U.P. Singh, A. Bali, S. Jain, Neuro-fuzzy elman wavelet control for nonlinear uncertain systems with fuzzy input and unknown fuzzy disturbances: application to robotics. Int. J. Adapt. Control Signal Process
27. H. Nguyen, T. Dang, K. Alexis, The reconfigurable aerial robotic chain: modeling and control, in *2020 IEEE International Conference on Robotics and Automation (ICRA)* (IEEE, 2020), pp. 5328–5334

How Viscosity of an Asphalt Binder Depends on Temperature: Theoretical Explanation of an Empirical Dependence

Edgar Daniel Rodriguez Velasquez and Vladik Kreinovich

Abstract Pavement must be adequate for all the temperatures, ranging from the winter cold to the summer heat. In particular, this means that for all possible temperatures, the viscosity of the asphalt binder must stay within the desired bounds. To predict how the designed pavement will behave under different temperatures, it is desirable to have a general idea of how viscosity changes with temperature. Pavement engineers have come up with an empirical approximate formula describing this change. However, since this formula is purely empirical, with no theoretical justification, practitioners are often somewhat reluctant to depend on this formula. In this paper, we provide a theoretical explanation for this empirical formula—namely, we should that this formula can be naturally derived from natural invariance requirements.

1 Formulation of the Problem

What is asphalt binder: a brief reminder. In most pavements, the top—asphalt—layer is formed by mixing the asphalt binder (which is usually the residue from petroleum refining) with rocks of appropriate sizes.

It is important to maintain the right viscosity of the asphalt binder. In general, if a stress is applied to a material, it first deforms and then, if the stress increases further, it cracks. Some materials are brittle, a reasonably small stress caused then to

E. D. Rodriguez Velasquez
Department of Civil Engineering, Universidad de Piura in Peru (UDEP), Av. Ramón Mugica 131, Piura, Peru
e-mail: edgar.rodriguez@udep.pe; edrodriguezvelasquez@miners.utep.edu

Department of Civil Engineering, University of Texas at El Paso, 500 W. University, El Paso, TX 79968, USA

V. Kreinovich (✉)
Department of Computer Science, University of Texas at El Paso, 500 W. University, El Paso, TX 79968, USA
e-mail: vladik@utep.edu

N. H. Phuong and V. Kreinovich (eds.), *Deep Learning and Other Soft Computing Techniques*, Studies in Computational Intelligence 1097,
https://doi.org/10.1007/978-3-031-29447-1_18

crack, other materials are more flexible, they undergo a reasonable large amount of deformation before they start cracking.

Materials like asphalt binders are, from the physical viewpoint, highly viscous liquids (such materials are also called *semi-solids*). Their mechanical properties are different from the properties of usual liquids like water because, in comparison to water, the asphalt binders have a very high viscosity. The asphalt binder's reaction to stress is determined by this viscosity, and it affects the quality of the resulting pavement:

- If the viscosity is too low, the pavement is deformed by traffic and this deformation will make the pavement non-smooth and thus, hinder the traffic.
- If the viscosity is too high, i.e., if the pavement is too stiff, then the stress caused by the traffic will cause cracks. As more traffic passes through the pavement segment, the cracks will grow, and the resulting large cracks will also hinder the traffic.

It is therefore important to maintain the proper value of viscosity.

How viscosity changes with temperature. For all liquids—and asphalt binders are not an exception—viscosity η decreases with temperature T. Since a pavement has to function properly at all seasons, during the wide range of outside temperatures, it is important to take this dependence into account when designing the pavement. Pavement engineering uses the following empirical formula [1, 2] to describe this dependence:

$$\ln\left(\ln\left(\frac{\eta}{\eta_0}\right)\right) = a + b \cdot \ln(T) \quad (1)$$

for some constants η_0, a, and b, where the temperature is measured in absolute units (e.g., Kelvins).

Formulation of the problem and what we do in this paper. The formula (1) is empirical. In general, empirical formulas are not as reliable as formulas that have at least some theoretical explanation—these explanations make the use of such formulas outside their original testing example more convincing. It is therefore desirable to come up with a theoretical explanation for this formula. Such an explanation is provided in this paper.

2 Our Explanation

Invariance: main idea behind our explanation. Our explanation is based on the fact that the numerical value of a physical quantity depends on the selection of the measuring unit—and sometimes, e.g., when we measure time, also on the selection of the starting point. If we replace the original measuring unit with a new unit which is λ times smaller, then each numerical value x changes to $\lambda \cdot x$: for example, 2 kg becomes $1000 \cdot 2 = 2000$ g. Similarly, if we replace the original starting point with the new one which is x_0 units lower, then each numerical value changes to $x + x_0$.

In many situations, there is no physically preferred measuring unit, selection of a measuring unit is simply a matter of convenience. In this case, it makes sense to assume that the exact form of the dependence $y = f(x)$ between the two quantities x and y should remain the same if we change the measuring unit for x. Of course, we may also need to appropriately change the measuring unit for y: e.g., the formula $y = x^2$ relating the area y of a square region to the length x of its side does not depend on what unit we use to measure length. However, to preserve this formula, when we change the unit of length from kilometers to miles, we also need to change the unit of area from square kilometers to square miles.

Invariance: precise definitions. So, in the case when we consider scaling of x and scaling of y, the corresponding *scale-scale* invariance of the relation $y = f(x)$ means that for every value $\lambda > 0$ there exists a value $\mu > 0$—in general, depending on λ—for which $y = f(x)$ implies that $Y = f(X)$, where $X = \lambda \cdot x$ and $Y = \mu \cdot y$. Similarly:

- *shift-scale* invariance of the relation $y = f(x)$ means that for every value x_0 there exists a value $\mu > 0$—in general, depending on x_0—for which $y = f(x)$ implies that $Y = f(X)$, where $X = x + x_0$ and $Y = \mu \cdot y$;
- *scale-shift* invariance of the relation $y = f(x)$ means that for every value $\lambda > 0$ there exists a value y_0—in general, depending on λ—for which $y = f(x)$ implies that $Y = f(X)$, where $X = \lambda \cdot x$ and $Y = y + y_0$; and
- *shift-shift* invariance of the relation $y = f(x)$ means that for every value x_0 there exists a value y_0—in general, depending on x_0—for which $y = f(x)$ implies that $Y = f(X)$, where $X = x + x_0$ and $Y = y + y_0$.

Which dependencies are invariant. For each of the above four types of invariance, we have a full description of all the functions which are correspondingly invariant (see, e.g., [3]):

- all scale-scale invariant functions have the form $f(x) = A \cdot x^a$ for some values A and a;
- all shift-scale invariant functions have the form $f(x) = A \cdot \exp(a \cdot x)$ for some values A and a;
- all scale-shift invariant functions have the form $f(x) = A \cdot \ln(x) + a$ for some values A and a; and
- all shift-shift invariant functions have the form $f(x) = A \cdot x + a$ for some values A and a.

Some dependencies are indirect. In many practical situations, while there is a dependence between quantities x and y, this dependence is not direct: the quantity x affects some other quantity z, and that quantity z affects y—and we may have an even longer chain of effects. In the case of the dependencies $x \to z \to y$, it is often reasonable to assume that both dependencies $z = g(x)$ and $y = h(z)$ are invariant. In this cases, the resulting dependence of y on x is a composition of two invariant functions $y = h(g(x))$; see [3] for examples.

What natural invariances we have in our case. In our case, both for viscosity and for temperature (when described ion absolute units), there is a definite starting point: 0 viscosity and absolute zero temperature. However, in both cases, there does not seem to be a preferred measuring unit. It is therefore reasonable to assume that for both quantities, the natural transformation is scaling—corresponding to the selection of a different measuring unit.

The observed dependence is not scale-scale invariant. Based on the natural invariances, one may expect that the dependence of the viscosity on temperature should be described by the scale-scale invariant power law $\eta = A \cdot T^a$. However, the empirical dependence (1) in different. Namely, if we apply the exponential function $\exp(z)$ to both sides of the formula (1), we get

$$\ln\left(\frac{\eta}{\eta_0}\right) = \exp(a + b \cdot \ln(T)) = \exp(a) \cdot \exp(b \cdot \ln(T)) =$$

$$\exp(a) \cdot (\exp(\ln(T))^b = A \cdot T^b, \tag{2}$$

where we denoted $A \stackrel{\text{def}}{=} \exp(a)$. By applying $\exp(z)$ to both sides of the equality (2), we conclude that

$$\frac{\eta}{\eta_0} = \exp(A \cdot T^b), \tag{3}$$

and thus, that

$$\eta = \eta_0 \cdot \exp(A \cdot T^b). \tag{4}$$

This is different from the power law.

In other words, the observed dependence of η on T cannot be explained by the direct invariance. Thus, a natural idea is to see if this empirical dependence can be explained as indirect dependence.

Indirect invariance indeed explains the empirical dependence. In line with the conclusion of the previous subsection, let us assume that η depends on some auxiliary quantity z that, in its turn, depends on T, i.e., that the dependence $\eta = f(T)$ has the form $\eta = h(g(T))$, i.e., the form $\eta = h(z)$ and $z = g(T)$.

We know that for both T and η, natural transformations are scalings. So, depending on what transformations we assume for z in both dependencies, we get the following four possible cases:

1. the dependence of z on $Y = T$ is scale-scale-invariant, and the dependence of η on z is also scale-scale-invariant;
2. the dependence of z on $Y = T$ is scale-shift-invariant, and the dependence of η on z is shift-scale-invariant;
3. the dependence of z on $Y = T$ is scale-scale-invariant, and the dependence of η on z is shift-scale-invariant;
4. the dependence of z on $Y = T$ is scale-shift-invariant, and the dependence of η on z is scale-scale-invariant.

Let us analyze these cases one by one.

1. In the first case, we have $z = A_1 \cdot T^{a_1}$ for some A_1 and a_1, and $\eta = A_2 \cdot z^{a_2}$. Substituting the expression for z into the formula describing the dependence of η on z, we conclude that

$$\eta = A_2 \cdot (A_1 \cdot T^{a_1})^{a_2} = (A_2 \cdot A_1^{a_2}) \cdot T^{a_1 \cdot a_2}, \tag{5}$$

i.e., $\eta = A \cdot T^a$, where we denoted $A \stackrel{\text{def}}{=} A_2 \cdot A_1^{a_2}$ and $a \stackrel{\text{def}}{=} a_1 \cdot a_2$. Thus, we get a power law—and we have already mentioned that the actual dependence is different from the power law.
2. In the second case, we have $z = A_1 \cdot \ln(T) + a_1$ for some A_1 and a_1, and $\eta = A_2 \cdot \exp(a_2 \cdot z)$. Substituting the expression for z into the formula describing the dependence of η on z, we conclude that

$$\eta = A_2 \cdot \exp(a_2 \cdot (A_1 \cdot \ln(T) + a_1)) = A_2 \cdot \exp(a_2 \cdot A_1 \cdot \ln(T)) \cdot \exp(a_2 \cdot a_1) =$$

$$(A_2 \cdot \exp(a_2 \cdot a_1)) \cdot (\exp(\ln(T))^{a_2 \cdot A_1} = (A_2 \cdot \exp(a_2 \cdot a_1)) \cdot T^{a_2 \cdot A_1}, \tag{6}$$

i.e., $\eta = A \cdot T^a$, where we denoted $A \stackrel{\text{def}}{=} A_2 \cdot \exp(a_2 \cdot a_1)$ and $a \stackrel{\text{def}}{=} a_2 \cdot A_1$. Thus, we also get a power law.
3. In the third case, we have $z = A_1 \cdot T^{a_1}$ for some A_1 and a_1, and $\eta = A_2 \cdot \exp(a_2 \cdot z)$. Substituting the expression for z into the formula describing the dependence of η on z, we conclude that

$$\eta = A_2 \cdot \exp(a_2 \cdot A_1 \cdot T^{a_1}) = A_2 \cdot \exp((a_2 \cdot A_1) \cdot T^{a_1}), \tag{7}$$

i.e., we get the formula (4) for $\eta_0 = A_2$, $A = a_2 \cdot A_1$, and $b = a_1$.
4. In the fourth case, we have $z = A_1 \cdot \ln(T) + a_1$ for some A_1 and a_1, and $\eta = A_2 \cdot z^{a_2}$. Substituting the expression for z into the formula describing the dependence of η on z, we conclude that

$$\eta = A_2 \cdot (A_1 \cdot \ln(T) + a_1)^{a_2}. \tag{8}$$

At first glance, this expression does not look like the desired formula (4), but it actually describes, under this formula, the dependence of T on η. Indeed, in this case, taking logarithm of both sides of the formula (3), we get

$$\ln(\eta) - \ln(\eta_0) = A \cdot T^a,$$

hence

$$T^a = A^{-1} \cdot \ln(\eta) - A^{-1} \cdot \ln(\eta_0),$$

and

$$T = (A^{-1} \cdot \ln(\eta) - A^{-1} \cdot \ln(\eta_0))^{1/a}.$$

This formula can be described as

$$T = A_2 \cdot (A_1 \cdot \ln(\eta) + a_1)^{a_2}, \quad (8a)$$

if we take $A_2 = 1$, $A_1 = A^{-1}$, $a_1 = -A^{-1} \cdot \ln(\eta_0)$, and $a_2 = 1/a$.

So, indeed, indirect invariance explains the desired formula—to be more precise, it explains either the dependence of η on T or the dependence of T on η. Which of the two formulas (7) and (8) should be applied to the dependence of η on T should be determined experimentally, just like the numerical values of all the corresponding parameters.

Acknowledgements This work was supported in part by the National Science Foundation grants 1623190 (A Model of Change for Preparing a New Generation for Professional Practice in Computer Science), and HRD-1834620 and HRD-2034030 (CAHSI Includes), and by the AT&T Fellowship in Information Technology. It was also supported by the program of the development of the Scientific-Educational Mathematical Center of Volga Federal District No. 075-02-2020-1478, and by a grant from the Hungarian National Research, Development and Innovation Office (NRDI).

References

1. American Society for Testing and Materials (ASTM), *Standard Practice for Viscosity-Temperature Chart for Asphalt Binders* (ASTM International, International Standard ASTM D2493/D2493M-16, West Conshohocken, Pennsylvania, 2016)
2. S.W. Haider, M.W. Mirza, A.K. Thottempudi, J. Bari, G.Y. Baladi, Effect of test methods on viscosity temperature susceptibility characterization of asphalt binders for the mechanistic-empirical pavement design guide, in *Proceedings of the First Congress of the Transportation and Development Institute (T&DI) of the American Society of Civil Engineers (ASCE)* (Chicago, Illinois, 2011). 13–16 Mar 2011
3. E.D. Rodriguez Velasquez, V. Kreinovich, O. Kosheleva, Invariance-based approach: general methods and pavement engineering case study. Int. J. General Syst. **50**(6), 672–702 (2021)

Applications to Physics, Including Physics Behind Computations

Why in MOND—Alternative Gravitation Theory—A Specific Formula Works the Best: Complexity-Based Explanation

Olga Kosheleva and Vladik Kreinovich

Abstract Based on the rotation of the stars around a galaxy center, one can estimate the corresponding gravitational acceleration—which turns out to be much larger than what Newton's theory predicts based on the masses of all visible objects. The majority of physicists believe that this discrepancy indicates the presence of "dark" matter, but this idea has some unsolved problems. An alternative idea—known as Modified Newtonian Dynamics (MOND, for short) is that for galaxy-size distances, Newton's gravitation theory needs to be modified. One of the most effective versions of this idea uses so-called *simple interpolating function*. In this paper, we provide a possible explanation for this version's effectiveness. This explanation is based on the physicists' belief that out of all possible theories consistent with observations, the true theory is the simplest one. In line with this idea, we prove that among all the modifications which explain both the usual Newton's theory for usual distance and the observed interactions for larger distances, this so-called "simple interpolating function" is indeed the simplest—namely, it has the smallest computational complexity.

1 Formulation of the Problem

What started the whole thing: discrepancy between Newton's gravity, visible masses, and observations. According to Newton's theory of gravitation, bodies with masses m and M attract each other with the force

O. Kosheleva
Department of Teacher Education, University of Texas at El Paso, 500 W. University, El Paso, Texas 79968, USA
e-mail: olgak@utep.edu

V. Kreinovich (✉)
Department of Computer Science, University of Texas at El Paso, 500 W. University, El Paso, Texas 79968, USA
e-mail: vladik@utep.edu

N. H. Phuong and V. Kreinovich (eds.), *Deep Learning and Other Soft Computing Techniques*, Studies in Computational Intelligence 1097,
https://doi.org/10.1007/978-3-031-29447-1_19

$$F = \frac{G \cdot m \cdot M}{r^2}, \tag{1}$$

where G is a constant and r is the distance between the bodies. In accordance to Newton's Second Law, this force leads to an acceleration a of the mass-m body such that

$$F = m \cdot a. \tag{2}$$

Newton's theory of gravitation provides a very accurate description of all the motions in the Solar system: suffice it to say that the difference between Newton's theory and observations of Mercury's position—the difference that was one of the reasons for replacing this theory with a more accurate General Relativity Theory—amounted to less than a half angular second per year; see, e.g., [1–3].

In this application of Newton's theory, masses can be determined by observing the motion of celestial bodies. Specifically, if a body of mass m rotates with velocity v around a body of mass M at a distance r, then its acceleration is equal to

$$a = \frac{v^2}{r}. \tag{3}$$

By equating the force (2) corresponding to this acceleration with the gravitational force (1), we conclude that

$$\frac{m \cdot v^2}{R} = \frac{G \cdot m \cdot M}{r^2}, \tag{3}$$

thus we can compute the mass M of the central body as

$$M = \frac{v^2 \cdot R}{G}. \tag{4}$$

From the formula (3), we can also conclude that

$$v^2 = \frac{G \cdot M}{r}, \tag{5}$$

i.e., that for objects rotating around the same central body at different distances r, the square of the velocity is proportional to $1/r$.

Since Newton's theory works so well in the Solar system, a natural idea is to use it to describe other celestial bodies—e.g., rotation of stars in a stellar cluster or in a galaxy. Surprisingly, it turned out that, contrary to the formula (5), when we get to the faraway area, where there are very few visible sources, the velocity stays the same—and does not decrease with r; see, e.g., [4].

The usual explanation is that, in addition to the visible objects, the Universe contains a large amount of cold matter that practically does not emit radiation and is, thus, not visible to our telescopes. This *dark matter* is what most physicists believe in.

MOND—an alternative theory of gravitation. While the presence of dark matter explains a lot of empirical facts, there are also some serious issues with this idea. One of the main issues is that the main idea behind dark matter is that it should be reasonably independent from the usual matter, there should be very little correlation between the usual matter and the dark matter. However, the actual values of the dark matter mass obtained from the observed velocities seem to be well correlated with the amount of usual (visible) mass.

This correlation led some physicists to believe that there is no such thing as dark matter, but that, instead, Newton's theory needs to be modified. One of the ideas—that got some observable confirmations—is called Modified Newton's Dynamics, MOND, for short. In this theory, instead of the usual Newton's Second Law (2), we have a modified Newton's law

$$F = m \cdot g(a), \tag{6}$$

for an appropriate function $g(a)$ [5–7]. This theory is consistent with several other astrophysical phenomena; see, e.g., [8] and references therein.

Comment. To be precise, MOND uses an equivalent but slightly different formula $F = m \cdot a \cdot f(a)$ for some function $f(a)$. This is equivalent to our formula (6) if we take $g(a) = a \cdot f(a)$.

Specifics of MOND. In the Solar system, Newton's theory works well, So, it is reasonable to conclude that for reasonable-size accelerations a, we should have $g(a)$ asymptotically equivalent to a. On the other hand, in cluster- and galaxy-size range, where the distances are much larger and accelerations are much smaller, the force is proportional to r^{-2}, while the observed acceleration (3) decreases as r^{-1}. Thus, in this case, the gravitational force is proportional to a^2, so we should gave $g(a)$ asymptotically equivalent to $c \cdot a^2$ for some constant c. So, it is reasonable to impose the following requirement on $g(a)$.

Definition 1 *We say that the functions $f(x)$ and $g(x)$ are asymptotically equivalent when x tends to x_0 when*

$$\lim_{xc \to x_0} \frac{f(x)}{g(x)} = 1.$$

Definition 2 *We say that a function $g(a)$ is consistent with observations if:*

- *this function is asymptotically equivalent to a when $a \to \infty$, and*
- *this function is asymptotically equivalent to $c \cdot a^2$ (for some constant c) when $a \to 0$.*

Several such functions have been proposed. At present, one of the most effective is the function

$$f(a) = \frac{a^2}{a + a_0} \tag{7}$$

known as *simple interpolating function*.

Formulation of the problem. How can we explain why, out of all functions that satisfy Definition 2, namely the function (7) is the most efficient?

What we do in this paper. In this paper, we provide a possible explanation for this effectiveness. This explanation is based on the physicists' belief that our of all possible physical theories consistent with observations, the simplest one is the true one [1, 3]. In line with this idea, we show that of all the functions that satisfy Definition 2, the function (7) is the simplest—in the sense that it has the smallest computational complexity.

2 Our Explanation

How can we define computational complexity. In a computer, only arithmetic operations are handware supported, all other computations are reduced to a sequence of arithmetic operations. For example, if we ask a computer to compute $\exp(x)$, what it will actually do is compute the sum of the first few terms of the corresponding Taylor series, i.e., the expression

$$\exp(x) \approx 1 + x + \frac{x^2}{2} + \cdots + \frac{x^N}{N!}.$$

Alternatively, it may compute an approximating fractional-linear expression. In all these cases, each computation consists of several computational steps, each of which is addition $+$, subtraction $-$, multiplication $\times$, or division $/$.

Out of these operations:

- addition and subtraction are the fastest,
- multiplication takes longer time—since multiplication is, in effect, a sequence of additions, and
- division takes the longest time—since the usual way to perform division is to perform several multiplications.

Let us denote the times needed for each arithmetic operation as, correspondingly, $t_{\pm}$, $t_{\times}$, and $t_{/}$. Then, we have

$$t_{\pm} < t_{\times} < t_{/}. \tag{8}$$

Definition 3 *By a sequence of arithmetic operations, we means a finite sequence of triples $T_1, \ldots, T_n$, where each triple t has the form $\langle op, v_1, v_2\rangle$, where:*

- *op is equal to $+$, $-$, $\times$, or $/$, and*
- *for each tuple T_i, each v_j is:*
  - *either a symbol x_j, where j is a positive integer smaller than i,*
  - *or a real number,*
  - *or the original variable a.*

By the result *of applying a sequence to a number a, we mean the value of the last of numbers $x_1, \ldots, x_n$, where each x_i is the result of applying the corresponding operation op to the values v_1 and v_2.*

Example

- The expression $(a + 1) \cdot (a + 2)$ can be computed by the following sequence of arithmetic operations: $T_1 = \langle +, a, 1\rangle, T_2 = \langle +, a, 2\rangle$, and $T_3 = \langle \times, x_1, x_2\rangle$. In this case, $x_1 = a + 1, x_2 = a + 2$, and $x_3 = x_1 \cdot x_2$.
- The expression (7) can be computed by the following sequence of arithmetic operations: $T_1 = \langle \times, a, a\rangle$, $T_2 = \langle +, a, a_0\rangle$, and $T_3 = \langle /, x_1, x_2\rangle$. In this case, $x_1 = a \cdot a = a^2, x_2 = a + a_0$, and $x_3 = x_1/x_2 = a^2/(a + a_0)$.

Definition 4 *Let t, $t_\times$, and $t_/$ be positive real numbers for which $t_\pm < t_\times < t_/$. By the* computational complexity *of a sequence of arithmetic operations, we mean the sum of the numbers t_{op} corresponding to these operations.*

Example

- Computation of $(a + 1) \cdot (a + 2)$ consists of two additions and one multiplication, so the complexity is $2t_\pm + t_\times$.
- Computation of the expression (7) consists of one addition (to compute $a + a_0$), one multiplication (to compute $a^2 = a \cdot a$), and one division, so the complexity is $t_\pm + t_\times + t_/$.

Proposition *Out of all functions that are consistent with observations (in the sense of Definition 2), the function (7) has the smallest computational complexity.*

Comment. It should be mentioned that this result holds no matter what values $t_\pm < t_\times < t_/$ we select.

Proof

1°. Let us first us first prove that a function $g(a)$ that satisfies Definition 2 must contain at least one division.

Indeed, otherwise, it would be a polynomial, and the only way a polynomial is asymptotically equivalent to a for $a \to \infty$ is when it is a linear polynomial—otherwise, the highest power of a will dominate for large a. However, a linear polynomial cannot be asymptotically equivalent to $c \cdot a^2$ when $a \to 0$.

2°. Let us prove that the expression $g(a)$ cannot consist of only divisions and multiplications—and thus, it must contain at leats one addition or subtraction.

Indeed, in the beginning, all we have is a and constants c. Both inputs are expressions of the type $c \cdot a^d$ for some c and integer d: $a = 1 \cdot a^1$ and $c = c \cdot a^0$. One can easily check that if we multiply or divide such expressions, we will still have an expression of this type. However, no expression of this type satisfies Definition 2:

- the first condition from this definition implies that $d = 1$, while
- the second condition is only satisfied when $d = 2$.

3°. So, the desired expression must contain at least one division and at least one addition of subtraction. Depending on how many divisions and multiplications are used, we can consider the following three cases:

- The first case is when this expression contains two (or more) divisions.
- The second case is when this expression contains a single division, and all other operations are additions or subtractions.
- The remaining third case is when this expression contains a single division and at least one multiplication.

Let us consider these cases one by one.

3.1°. Let us first consider the first case, when the expression contains two (or more) divisions.

As we have shown in Part 2 of this proof, the expression must contain at least one addition of subtraction. Thus, the overall computational complexity must be larger than or equal to $2t_/ + t_\pm$. Since $t_/ > t_\times$, this value is larger than the value $t_/ + t_\times + t_\pm$ corresponding to (7). Since the expression (7) satisfies Definition 2, this means that the expression that contains two or more divisions cannot have the smallest computational complexity among all expressions that satisfy this definition.

3.2°. Let us now consider the second case, when the expression contains a single division, and when all other operations are additions or subtractions.

If we simply apply addition and subtraction to a and constants, all we get is expressions of the type $n \cdot a + c$, for some constants n and c. So, all we get after dividing such expressions is an expression of the type

$$\frac{n \cdot a + c}{n' \cdot a + c'},$$

and all we get by further additions are expressions of the type

$$\frac{n \cdot a + c}{n' \cdot a + c'} + n'' \cdot a + c''.$$

If we bring this sum to the common denominator, we get a fraction in which the numerator is quadratic in a and the denominator is linear in a:

$$\frac{n_2 \cdot a^2 + n_1 \cdot a + n_0}{d_1 \cdot a + d_0},$$

for some coefficients n_i and d_i.

If d_1 was equal to 0, then this expression could have been obtained without using division, and we have already shown, in Part 1 of this proof, that this is not possible. Thus, $d_1 \neq 0$. So, we can divide both the numerator and denominator by d_1 and get a simpler expression

$$\frac{N_2 \cdot a^2 + N_1 \cdot a + N_0}{a + a_0}, \text{ where } N_i \stackrel{\text{def}}{=} \frac{n_i}{d_1} \text{ and } a_0 \stackrel{\text{def}}{=} \frac{d_0}{d_1}.$$

If N_2 was equal to 0, then for $a \to \infty$, this expression would be a constant, but, according to Definition 2, it should grow as a. Thus, $N_2 \neq 0$, and for $a \to \infty$, this expression grows as $N_2 \cdot a$. Due to Definition 2, this implies that $N_2 = 1$, so we can further simplify this expression, into

$$\frac{a^2 + N_1 \cdot a + N_0}{a + a_0}.$$

If a_0 was equal to 0, then this expression would be equal to

$$a + N_1 + \frac{N_0}{a},$$

that does not provide the right a^2 asymptotic. Thus, $a_0 \neq 0$. So, for $a \to 0$, the denominator is asymptotically a constant, and thus, the corresponding asymptotic behavior of the ratio is propotional to the numerator.

If $N_0 \neq 0$, then:

- this expression would be tending to a non-zero constant when $a \to 0$,
- while we should have $\sim c \cdot a^2$.

Thus, we have $N_0 = 0$.

In this case, if we had $N_1 \neq 0$, then for $a \to 0$:

- this expression will be asymptotically linear,
- while we assumed it to be quadratic.

Thus, we must have $N_1 = 0$ as well.

In this case, the above expression takes the desired form (7).

3.3°. Finally, let us consider the case when, in addition to division and addition/subtraction, we also have at least one multiplication.

The expression (7) satisfies Definition 2 and contains exactly one of each operations, so it clearly has the smallest computational complexity of all such expressions. So, to complete the proof, we need to show that no other combination of these three operations leads to a function that satisfies Definition 2. Let us consider all such combinations.

4°. Similarly to Part 3.2, we can show that we cannot get any different function $g(a)$ if we simply add multiplication by a constant. Thus, we need at least one multiplication of expressions that contain a.

5°. Let us prove that for functions $g(a)$ that satisfy Definition 2, division cannot be the first operation.

Indeed, in the beginning, all we have is the variable a and constants. So, as a result of division, we get one of the following expressions:

- either divide a by itself, resulting in 1;
- or divide a by a constant c, resulting in $c^{-1} \cdot a$;
- or divide a constant by another constant—resulting in a new constant;
- or divide a constant by a, resulting in $c \cdot a^{-1}$.

In all these cases, we have an expression of the type $c \cdot a^n$ for some integer n. When we add and multiply such expression, all we have is a polynomial in terms of a and a^{-1}, i.e., a linear combination of the terms of this type. We can sort these terms in the increasing order of power and get

$$g(a) = c_1 \cdot a^{n_1} + \cdots + c_m \cdot a^{n_m} \text{ for } n_1 < \cdots < n_m.$$

Here:

- for $a \to \infty$, this expression is asymptotically equivalent to $c_m \cdot a^{n_m}$, so we must have $n_m = 1$,
- but for $a \to 0$, this expression is asymptotically equivalent to $c_1 \cdot a^{n_1}$, so we must have $n_1 = 2$.

This contradicts to the fact that $n_1 < n_m$. The statement is proven.

6°. Let us prove that for 3-step computations consisting of addition (or subtraction), multiplication, and division, addition/subtraction cannot be the last operation.

Indeed, based on Part 2 of this proof, as a result of division and multiplication, we can only get expressions of the type $c \cdot a^n$ for some integer n. If we apply addition ot subtraction, we get a linear combination of such expressions, and we have shown, in Part 5 of this proof, that such linear combination cannot satisfy Definition 2.

7°. Now we know that in the currently considered third case, when the expression contains a single division and at least one multiplication, the fastest computation scheme consists of 3 operations: addition or subtraction, multiplication, and division.

We also know that:

- addition/subtraction cannot be the last operation, so it must be first or second, and
- division cannot be the first operation, so it must be second or third.

In view of these two requirements, let us analyze what are the possible orders of these three operations.

- If addition/subtraction is the first, then division can be second or third, and multiplication must be, correspondingly, the third or the second one. So, in this case, we can have the following sequence of operations: $\pm$, $/$, $\times$ or $+$, $\times$, $/$.
- If addition/subtraction is the second operation, then division cannot be second, it must be third, and the only remaining place for multiplication is to be the first operation. In this case, we have the following sequence of operations: $\times$, $\pm$, $/$.

Let us consider the resulting three possible orders one by one.

8°. Let us first consider the order $\pm$, $/$, $\times$.

For this order, for the first operation, we start with a and constant. Thus, we have two possible non-trivial options for the first operation:

- either add or subtract a and a constant, resulting in $c \pm a$, or
- add two a's, resulting in $2a$.

The remaining two possible version of the first operations do not lead to any non-trivial results. Indeed:

- If we add two constants, we get a new constant—so this step is not needed and cannot be part of the shortest computation.
- Similarly, if we subtract a from a, we get a constant 0, which also cannot be a part of the shortest computation.

In both non-trivial operations, after the first step, we get expressions that are linear in terms of a.

After the first step, we divide two expressions A/B, and then perform multiplication. This multiplication must include the result of division—otherwise, this result is not used and we could get a faster computation scheme by skipping this step. So, we:

- either multiply A/B by some other expression C,
- or multiply the expression A/B by itself.

Let us consider these two options one by one.

8.1°. Let us first consider the option when we multiply A/B by C.

For this option, we get $(A/B) \cdot C$, which is equivalent to $(A \cdot C)/B$, i.e., we get a quadratic expression divided by a linear expression. We have already shown—in Part 3.2 of this proof—that in this case, Definition 2 leads to the expression (7).

8.2°. Let us now consider the option when we multiply the fractional-linear expression A/B by itself.

If the denominator was a constant, then we would not need division, and we know, from Part 1 of this proof, that this is not possible. Thus, the denominator is a linear function. In this case, when $a \to \infty$, then A/B tends to a constant. Thus, the product $(A/B)^2$ cannot be asymptotically equivalent to a. So, in this case, we do not get an expression that satisfies Definition 2.

9°. Let us now consider the order $\pm$, $\times$, $/$.

In the first step, we get an expression which is linear in a. After the multiplication step, we get an expression which is quadratic in a. We must use this quadratic expression in division—otherwise, multiplication would not be needed at all, and we have already considered this case. There are thus two options here:

- We can divide the quadratic expression by a linear expression. Then, as we have already shown in Part 3.2 of this proof, we get the expression (7).
- Alternatively, we can divide a linear expression by a quadratic expression. Then the ratio cannot grow as a when $a \to \infty$.

So, in this case, we also get only the expression (7).

10°. Finally, let us consider the remaining order $\times$, $\pm$, $/$.

For this order, we first multiply. We start with a and constants. Thus, we get the following options:

- We can multiply a constant by a constant—but this will simply lead to a new constant, that we could have gotten from the very beginning. So, this step cannot be part of the computation with the smallest computational complexity.
- We can multiply a by a constant, resulting in $c \cdot a$. In this case, on the second $\pm$ step, we will get linear functions of a, so the ratio obtained on the third computational step will be fractionally linear. And we have already shown that fractionally linear functions do not satisfy Definition 2.
- The only remaining option is to multiply a by a, resulting in a^2. In this option, as a result of $\pm$, we get quadratic functions, so as a result of division, we get either ratio of quadratic and linear function—which we already know leads to (7), or a ratio of two quadratic functions which tends to a constant when $a \to \infty$ and thus, also does not satisfy Definition 2.

So, the only possible case is the expression (7).

11°. So, we have shown that in all possible cases, the only expression $g(a)$ with the smallest computational complexity is indeed the expression (7). The proposition is proven.

Acknowledgements This work was supported in part by the National Science Foundation grants 1623190 (A Model of Change for Preparing a New Generation for Professional Practice in Computer Science), and HRD-1834620 and HRD-2034030 (CAHSI Includes), and by the AT&T Fellowship in Information Technology. It was also supported by the program of the development of the Scientific-Educational Mathematical Center of Volga Federal District No. 075-02-2020-1478, and by a grant from the Hungarian National Research, Development and Innovation Office (NRDI).

References

1. R. Feynman, R. Leighton, M. Sands, *The Feynman Lectures on Physics* (Addison Wesley, Boston, Massachusetts, 2005)
2. C.W. Misner, K.S. Thorne, J.A. Wheeler, *Gravitation* (W. H. Freeman, New York, 1973)
3. K.S. Thorne, R.D. Blandford, *Modern Classical Physics: Optics, Fluids, Plasmas, Elasticity, Relativity, and Statistical Physics* (Princeton University Press, Princeton, New Jersey, 2017)
4. V.C. Rubin, W.K. Ford Jr., Rotation of the Andromeda Nebula from a spectroscopic survey of emission regions. Astrophys J **159**, 379–403 (1970). https://doi.org/10.1086/150317
5. M. Milgrom, A modification of the Newtonian dynamics as a possible alternative to the hidden mass hypothesis. Astrophys. J. **270**, 365–370 (1983). https://doi.org/10.1086/161130
6. M. Milgrom, A modification of the Newtonian dynamics-Implications for galaxies. Astrophys. J. **270**, 371–383 (1983). https://doi.org/10.1086/161131
7. M. Milgrom, A modification of the Newtonian dynamics-Implications for galaxy systems. Astrophys. J. **270**, 384–389 (1983). https://doi.org/10.1086/161132
8. P. Kroupa, T. Jerabkova, I. Thies, J. Pflamm-Altenburg, B. Famaey, H. Boffin, J. Dabringhausen, G. Beccari, T. Prusti, C. Boily, H. Haghi, X. Wu, J. Haas, A.H. Zonoozi, G. Thomas, L. Šubr, S.J. Aarseth, Asymmetrical tidal tails of open star clusters: stars crossing their cluster's práh challenge Newtonian gravitation. Mon. Not. R. Astron. Soc. **517**(3), 3613–3639 (2022). https://doi.org/10.1093/mnras/stac2563

Why Color Optical Computing?

Victor L. Timchenko, Yury P. Kondratenko, and Vladik Kreinovich

Abstract In this paper, we show that requirements that computations be fast and noise-resistant naturally lead to what we call color-based optical computing.

1 Why Optical Computing in the First Place: We Want Fast Computing

We want computers to be fast. One of the main reasons why computers were invented in the first place is that they can perform computations fast, much faster than humans. As a result, during the same time, computers can perform much more computations—and can, thus, solve computational problems that cannot be solved without them.

We want computers to be faster. Modern computers are very fast—but still, there are many important practical problems, to solve which we need faster computational devices. For example, nowadays, we can reasonably well predict tomorrow's weather—by using several hours of high-performance computers. In principle, similar computations can help us determine, in reasonable accuracy, in what direction a deadly tornado will turn in the next 15 min—but on modern computers, this computation also takes several hours, way after the tornado already moved. Because of this

V. L. Timchenko
Admiral Makarov National University of Shipbuilding, 9 Heroes of Ukraine Avenue, Mykolaiv 054025, Ukraine

Y. P. Kondratenko
Petro Mohyla Black Sea National University, 10 68th Desantnykiv Str., Mykolaiv 054003, Ukraine

V. Kreinovich (✉)
Department of Computer Science, University of Texas at El Paso, 500 W. University, El Paso, Tx 79968, USA
e-mail: vladik@utep.edu

N. H. Phuong and V. Kreinovich (eds.), *Deep Learning and Other Soft Computing Techniques*, Studies in Computational Intelligence 1097,
https://doi.org/10.1007/978-3-031-29447-1_20

and similar challenges, engineers and scientists try to come up with faster and faster computational devices.

The most time-consuming part of computations is communication. In modern computers, most time is spent not on computations themselves, but rather on communicating information between different part of the computer; see, e.g., [3]. For example, in the usual computer, the time needed to fetch a number from a memory is much larger than the time needed to perform an arithmetic operation—this is why compilers try to guess which values will be used in future operations and pre-fetch them by moving them from the general memory to registers—the fastest-to-access part of computer memory.

Maximizing communication speed leads to optical computing. Because of this, one of the main ideas of how to make computers faster is to make these communications as fast as possible. In nature, all speeds are limited by the speed of light; see, e.g., [2, 13]. Thus, a natural way to make computers faster is to make sure that all communications are performed with the largest possible speed—i.e., that they are performed by light (or by other types of electromagnetic waves). So, we naturally arrive at the idea of optical computing.

Optical computing is already successful. The idea of using optical processes in computing is not new, and it has been successfully applied; see, e.g., [5]. For example, optical processes have been successfully used to perform Fourier transform of images—and important step in many image processing algorithms; see, e.g., [6].

2 Why Color Optical Computing: Need for Robustness

We want robustness. Speed is not the only thing what we want from a computational device. We also want our computations to be reliable, to be accurate in spite of the inevitable noise—in other words, we want computations to be robust.

This means, in particular, that communications should be robust. How can we achieve this?

How can we achieve robustness: analysis of the problem. There are many factors affecting electromagnetic waves. Most of these factors are stationary—at least during the micro-times during the signal is transmitted, the properties of the media do not change much. Another important fact is that Maxwell's equations—that describe the electromagnetic waves—are linear; see, e.g., [2, 13]. So, all signal distortions are linear.

A signal can be described by describing how the signal's intensity $x(s)$ changes with time s. In general, the medium distorts the signal. So, instead of the original signal $x(s)$ we now have a somewhat different (distorted) signal whose value $y(t)$, at each moment t, is, in general, different from $x(t)$. This distortion is described by Maxwell's equations which are linear. Thus, the formula that describes the distorted signal in terms of the original signal is also linear.

A general linear dependence of a quantity y on finitely many variables $x_1, \ldots, x_n$ has the form

$$y = a + \sum_{i=1}^{n} b_i \cdot x_i$$

for some constants a and b_i. To describe a general linear dependence of a quantity y on infinitely many variables $x(s)$ corresponding to different moments of time s, we need to replace the finite sum with its infinite analogue—the integral, so we get

$$y = a + \int b(s) \cdot x(s)\, ds$$

for some values a and $b(s)$, where s is a continuous analogue of the index i.

In our case, we want to describe how for each moment of time t, the value $y(t)$ of the distorted signal y at this moment t depends on the values $x(s)$. Each of these dependencies is described by the above linear formula. Of course, for different t, we need, in general, to use different values a and $b(s)$. Let us denote the coefficients a and $b(s)$ describing the dependence of $y(t)$ on $x(s)$ by $a(t)$ and $b(t, s)$. In terms of these notations, the formula describing the transformation of the original signal $x(t)$ into the distorted signal $y(t)$ takes the following form:

$$y(t) = a(t) + \int b(t, s) \cdot x(s)\, ds. \quad (1)$$

Stationary means that this transformation does not change with time—i.e., that if we input the same signal $x(t)$ shifted in time, i.e., the signal $x(t + t_0)$, then we get the same output similarly shifted in time, i.e., we get

$$y(t + t_0) = a(t) + \int b(t, s) \cdot x(s + t_0)\, ds. \quad (2)$$

If we introduce the new variables $t' = t + t_0$ and $s' = s + t_0$ in terms of which $t = t' - t_0$ and $s = s' - t_0$, then Eq. (2) takes the form

$$y(t') = a(t' - t_0) + \int b(t' - t_0, s' - t_0) \cdot x(s')\, ds'. \quad (3)$$

Renaming the variable back to t and s, we get

$$y(t) = a(t - t_0) + \int b(t - t_0, s - t_0) \cdot x(s)\, ds. \quad (4)$$

The two linear transformations (1) and (4) are identical, so all the coefficients must coincide. Thus, we have $a(t) = a(t - t_0)$ for all t and t_0—which means that a is simply a constant, and

$$b(t - t_0, s - t_0) = b(t, s) \quad (5)$$

for all t, s, and t_0. In particular, for $t_0 = s$, we conclude that $b(t, s) = b(t - s, 0)$. So, if we denote $B(t) \stackrel{\text{def}}{=} b(t, 0)$, we get $b(t, s) = B(t - s)$. Thus, the formula (1) takes the form

$$y(t) = a + \int B(t - s) \cdot x(s)\, ds. \quad (6)$$

This is known as *convolution* of two functions $B(t)$ and $x(s)$. It is known (see, e.g., [10]) that the Fourier transform $\widehat{y}(\omega)$ of the convolution is equal to the product of the Fourier transforms of the convolved functions:

$$\widehat{y}(\omega) = \widehat{B}(\omega) \cdot \widehat{x}(\omega). \quad (7)$$

We do not know what are the values $\widehat{B}(\omega)$, the only thing we know is that for any random process, almost all these values are, with probability 1, different from 0. Thus, after an unknown transformation, while the actual values of Fourier transform are *not* preserved, the only thing that *is* preserved is whether each of the values is 0 or not.

So, we end up with the following idea.

Resulting idea of color optical computing. To represent a signal, we just indicate for which frequencies, the corresponding component is non-zero. In other words, different signals correspond to different sets of frequencies—i.e., to different subsets $S \subseteq (0, \infty)$ of the real line.

What is the physical meaning of this? Fourier transform means representing the original signal $x(t)$ as a linear combination of different sinusoids

$$x(t) = \int \widehat{x}(\omega) \cdot (\cos(\omega \cdot t) + \mathrm{i} \cdot \sin(\omega \cdot t))\, d\omega.$$

For light, different frequencies correspond to different pure colors. Thus, the corresponding signal simply corresponds to indicating which pure colors are present.

Because of this interpretation, it is natural to call this idea *color optical computing*.

Operations. If we combine together signals corresponding to sets S and S', then the combined signal corresponds to the set $S \cup S'$.

We can also suppress some frequencies by applying the corresponding filters. A filter can be described by the set S of all the frequencies that it keeps; all other frequencies it filters out. The set of all the frequencies that it filters out is the complement $-S = \{s : s \notin S\}$ to the set of all the frequencies that it retains. If we start with a signal corresponding to the set S, and we apply a filter that keeps the frequencies from some set S', then the resulting signal will contains all the frequencies from the set S that are also included in the set S'—i.e., it will correspond to the *intersection*

$$S \cap S' = \{s \in S : s \in S'\}$$

of these two sets.

Comment. Set operations naturally correspond to logical operations; for example:

- union corresponds to (inclusive) "or", in the sense that a frequency s belongs to the union $S \cup S'$ if and only if s belongs to the set S *or* s belongs to the set S';
- the complement corresponds to negation "not", in the sense that a frequency s belongs to the complement $-S'$ if and only if s does *not* belong to the set S';
- intersection corresponds to "and", in the sense that a frequency s belongs to the intersection $S \cup S'$ if and only if s belongs to the set S *and* s belongs to the set S'.

A natural example of color optical computing. If we start with the three basic colors—R (red), G (green), and B (blue)—then possible signals correspond to all $2^3 = 8$ possible subsets of the three-element sets:

- the empty set corresponding to black,
- the three fundamental colors themselves, corresponding to sets

$$\{R\}, \ \{G\}, \text{ and } \{B\},$$

- the three mixed colors $R + G$, $R + B$, and $G + B$ corresponding to sets

$$\{R, G\}, \ \{R, B\}, \text{ and } \{G, B\},$$

- and the white color—the mixture of all three basic colors—corresponding to the set

$$\{R, G, B\}.$$

Correspondingly, we can have filters that filter our some of the basic colors. Namely:

- We can have a filter that does not keep any of the fundamental colors at all. This filter corresponds to the empty set $\emptyset$.
- We can have filters that retain only one of the three fundamental colors. Depending on which fundamental color is retains, we get three filters correspond to the sets

$$\{R\}, \ \{G\}, \text{ and } \{B\}.$$

- We can also have filters that retain two of the three fundamental colors. Depending on which fundamental colors are retained, we get three filters correspond to the sets

$$\{R, G\}, \ \{R, B\}, \text{ and } \{G, B\}.$$

- We can also have a filter that filters out all three basic colors. This filter corresponds to the set

$$\{R, G, B\}.$$

For example:

- If we apply the red filter—i.e., filter corresponding to the set $S' = \{R\}$—to white light (that corresponds to the set $\{R, G, B\}$), we will get the set $S \cap S' = \{R\}$ corresponding to the red light.
- On the other hand, if we apply the same red filter to the light $S = \{B, G\}$ consisting of blue and green, we will get the empty set $S \cap S' = \emptyset$ corresponding to the black color.

Color optical computing based on the three basic colors can be effectively used in computations. The above example leads to a promising scheme [11, 12] for processing expert's uncertainty—as described in terms of fuzzy degrees (see, e.g., [1, 4, 7–9, 14]).

In particular, this idea can be used to ensure safety of ship navigation [12].

Acknowledgements This work was supported in part by the National Science Foundation grants 1623190 (A Model of Change for Preparing a New Generation for Professional Practice in Computer Science), and HRD-1834620 and HRD-2034030 (CAHSI Includes), and by the AT&T Fellowship in Information Technology.

It was also supported by the program of the development of the Scientific-Educational Mathematical Center of Volga Federal District No. 075-02-2020-1478, and by a grant from the Hungarian National Research, Development and Innovation Office (NRDI).

References

1. R. Belohlavek, J.W. Dauben, G.J. Klir, *Fuzzy Logic and Mathematics: A Historical Perspective* (Oxford University Press, New York, 2017)
2. R. Feynman, R. Leighton, M. Sands, *The Feynman Lectures on Physics* (Addison Wesley, Boston, 2005)
3. J.L. Hennessy, D.A. Patterson, *Computer Architecture: A Quantitative Approach* (Morgan Kaufmann, Cambridge, 2019)
4. G. Klir, B. Yuan, *Fuzzy Sets and Fuzzy Logic* (Prentice Hall, Upper Saddle River, 1995)
5. X. Li, Z. Shao, M. Zhu, J. Yang, *Fundamentals of Optical Computing Technology: Forward the Next Generation Supercomputer* (Springer, Singapore, 2018)
6. A.J. Macfaden, G.S.D. Gordon, T.D. Wilkinson, An optical Fourier transform coprocessor with direct phase determination. Sci. Rep. 7(1) Paper 13667 (2017). https://doi.org/10.1038/s41598-017-13733-1
7. J.M. Mendel, *Uncertain Rule-Based Fuzzy Systems: Introduction and New Directions* (Springer, Cham, 2017)
8. H.T. Nguyen, C.L. Walker, E.A. Walker, *A First Course in Fuzzy Logic* (Chapman and Hall/CRC, Boca Raton, 2019)
9. V. Novák, I. Perfilieva, J. Močkoř, *Mathematical Principles of Fuzzy Logic* (Kluwer, Boston, 1999)
10. B.G. Osgood, *Lectures on the Fourier Transform and Its Applications* (American Mathematical Society, Providence, 2019)
11. V. Timchenko, Y. Kondratenko, V. Kreinovich, Efficient optical approach to fuzzy data processing based on colors and light filter. Int. J. Probl. Control Inf. **52**(4) (2022)

12. V. Timchenko, Y. Kondratenko, V. Kreinovich, Decision support system for the safety of ship navigation based on optical color logic gates, in *Proceedings of the IX International Conference "Information Technology and Implementation" IT&I-2022*. Kyiv, Ukraine, November 30 – December 2 (2022)
13. K.S. Thorne, R.D. Blandford, *Modern Classical Physics: Optics, Fluids, Plasmas, Elasticity, Relativity, and Statistical Physics* (Princeton University Press, Princeton, 2017)
14. L.A. Zadeh, Fuzzy sets. Inf. Control **8**, 338–353 (1965)

Non-localized Physical Processes Can Help Speed Up Computations, Be It Hidden Variables in Quantum Physics or Non-localized Energy in General Relativity

Michael Zakharevich, Olga Kosheleva, and Vladik Kreinovich

Abstract While most physical processes are localized—in the sense that each event can only affect events in its close vicinity—many physicists believe that some processes are non-local. These beliefs range from more heretic—such as hidden variables in quantum physics—to more widely accepted, such as the non-local character of energy in General Relativity. In this paper, we attract attention to the fact that non-local processes bring in the possibility of drastically speeding up computations.

1 Localized Character of Physical Processes Limits Computation Speed

Most physicists believe that all processes are localized. According to modern physics, all speeds are limited by the speed of light c. This means, in particular, that all physical processes are *localized*—if some event is happening at a spatial location x at moment t, then at a next moment of time $t + \Delta t$, the only objects that can be affected by this event are located at distance $\le c \cdot \Delta t$ from the location x; see, e.g., [3, 8].

Localized character of physical processes limit computation speed. How does the localized character of physical processes affect computations? At any level of tech-

M. Zakharevich
SeeCure Systems, Inc., 7536 Seacoast Drive, Parkland, FL 33067, USA
e-mail: mzakharevich@gmail.com

O. Kosheleva
Department of Teacher Education, University of Texas at El Paso, 500 W. University, El Paso, TX 79968, USA
e-mail: olgak@utep.edu

V. Kreinovich (✉)
Department of Computer Science, University of Texas at El Paso, 500 W. University, El Paso, TX 79968, USA
e-mail: vladik@utep.edu

N. H. Phuong and V. Kreinovich (eds.), *Deep Learning and Other Soft Computing Techniques*, Studies in Computational Intelligence 1097,
https://doi.org/10.1007/978-3-031-29447-1_21

nological advancement, there are natural limitations on how faster a single processor can compute. To perform computations faster, a natural idea is to have several processors working in parallel. This is exactly how modern high-performance computers work: they consist of thousands of processors working in parallel.

At first glance, it may seem that the more processors we use, the faster the resulting computations. However, this first impression is wrong: limits on communication speed affects the resulting computation speed; see, e.g., [7].

Indeed, suppose that we ask a parallel computer to perform some computations. This computer performs the corresponding computations, and at some time t seconds after the task was given, delivers the result of this computation to us.

This computer may be huge, it may go all the way to the Moon, but, because of the limitations on the communication speed, the only processors that could affect the computation results are the ones which are located at a distance not exceeding $r = c \cdot t$ from our location. Processors located further away from our location may be trying their best, but cannot affect the computation result—since during the time t, any communication can only reach the distance $c \cdot t$.

So, all processors affecting the computation result are located inside the sphere of radius $r = c \cdot t$, with the center at our location. The volume of this inside of the sphere is

$$V = \frac{4}{3} \cdot \pi \cdot r^3 = \frac{4}{3} \cdot \pi \cdot c^3 \cdot t^3. \tag{1}$$

How many processors can fit into this volume? Let ΔV be the smallest possible volume of a single processor that can be attained at a current technological level. This means that the volume occupied by each processor is larger than or equal to ΔV. If N is the overall number of processors inside this area, this means that the overall volume of all the processors in this area is larger than or equal to $N \cdot \Delta V$. This volume cannot exceed the overall volume (1) of this area: $N \cdot \Delta V \leq V$. Thus, we get an upper bound on the number of processors:

$$N \leq \frac{V}{\Delta V} = \frac{4}{3} \cdot \frac{\pi \cdot c^3}{\Delta V} \cdot t^3, \tag{3}$$

i.e.,

$$N \leq C \cdot t^3, \tag{4}$$

where we denoted

$$C \stackrel{\text{def}}{=} \frac{4}{3} \cdot \frac{\pi \cdot c^3}{\Delta V}.$$

In principle, anything that can be computed in parallel on N computers can also be computed sequentially, on a single processor:

- First, we simulate, on the single processor, the first computation steps on all N processors—first we simulate the first step of the first processor, then the first

step of the second processor, etc. This simulation of a single step requires N computation steps of the single processor.
- Then, we simulate, on the single processor, the second computation steps on all N processors—first we simulate the second step of the first processor, then the second step of the second processor, etc. This simulation of a single step also requires N computation steps of the single processor, etc.

To perform each computational step of the parallel computer, we need N steps of the single processor. Thus, such simulation-based sequential computation requires time $N \cdot t$.

Let us denote by T the smallest amount of computation time needed to perform this computation on a sequential computer. Since a sequential simulation of a parallel computer is one of the ways to perform the corresponding computations on a sequential computer, we conclude that $T \leq N \cdot t$. So, using the inequality (4), we can conclude that

$$T \leq C \cdot t^3 \cdot t = C \cdot t^4. \tag{5}$$

This inequality, in its turn, implies that $t^4 \geq C^{-1} \cdot T$, i.e., that

$$t \geq C^{-1/4} \cdot T^{1/4} = \text{const} \cdot c^{-3/4} \cdot T^{1/4}. \tag{6}$$

This lower bound on the computation time of parallel computation does not depend on the number of processors—it is an absolute lower bound preventing us from having an unlimited computation speedup.

2 But Are All Physical Processes Localized?

Hidden variables in quantum physics: a brief historical overview. Quantum physics describes micro-processes. For many of these processes, at present, we can only make probabilistic predictions. For example, for radioactive decay:

- we cannot predict at what moment the radioactive atom will decay,
- but we can predict—with high accuracy—the corresponding probabilities, so that we can accurately predict which proportion of the atoms will decay by time t.

Such probabilistic character of predictions did not start with quantum physics: similar probabilistic character is exhibited in statistical physics. For example:

- we cannot predict in what direction a small particle will move in a liquid due to Brownian motion,
- but we can predict, for a large number of such particles, how many of them will be in a certain area after a given amount of time.

In statistical physics, the probabilistic character of predictions is caused by the fact when we do not know the initial positions and velocities of all the particles: the more accurately we measured these values, the more accurate our predictions. These initial positions and velocities can be called *hidden variables* – these variables have perfect physical sense, but they are "hidden" from us in the sense that usually, we do not know their values, since measuring them would be too complicated.

Naturally, when quantum physics appeared, many physicists – including Einstein himself—believed that the probabilistic character of quantum physics can also be explained by the existence of appropriate hidden variables; see, e.g., [2]. The corresponding theories did not became mainstream since they violated the localization ideas: in these theories, during time t, an event can affect other events located at distances larger than $c \cdot t$.

This non-localness was usually viewed as a limitation of the then proposed hidden-variable theories. The situation changed with the appearance of so-called Bell's inequalities paper [1], according to which if probabilities described by quantum physics are correct, then certain inequalities between these probabilities should be observed, inequalities that are not satisfied if hidden variables are localized. After the paper [1] appeared, two choices remained:

- either quantum physics is correct, then Bell's inequalities are satisfied, and only non-local hidden variables are possible,
- or quantum physics is only an approximation to reality, then Bell's inequalities are not satisfied, and local hidden variables are possible.

Later experiments confirmed that Bell's inequalities are true – for these results the 2022 Nobel Prize in Physics was awarded. Thus, we can definitely conclude that even if hidden variables exist, they should be non-localized.

While still not mainstream, non-localized hidden variable theories are being considered even now; see, for example, [9, 10], where a neural network-type model based on such hidden variables leads to a natural explanation of many physical equations and phenomena.

Another example of a non-local phenomenon: energy of the gravitational field. Gravitational forces can perform useful work—e.g., they are the main course of energy in hydroelectric power plants. So naturally, the gravitational field has energy. In Newton's physics, this energy is easy to describe—it can be described the same way as the energy of any other physical field.

In general, a physical theory is described by the so-called *Lagrangian* L, an expression whose value at a given space-time point x depends on values of the physical fields $\varphi(x, t)$ and their derivatives at this 4-point (x, t). The equations of this described theory come from the assumption that the so-called *action* $S \stackrel{\text{def}}{=} \int L(x, t)\, d^3x\, dt$ attains the smallest possible value.

The problem of a finding a function (or functions) that minimizes a functional is known as a *variational optimization* problem; see, e.g., [3, 4]. Such problems are generalizations of the usual calculus-related optimization problems in which we want to find the values of the variables $x_1, \ldots, x_n$ for which a given objective function $f(x_1, \ldots, x_n)$ attains its smallest possible value. According to calculus, at each such point $(x_1, \ldots, x_n)$, all the partial derivatives of the function f are equal to 0:

$$\frac{\partial f}{\partial x_i} = 0.$$

Similarly, to find the function(s) $\varphi(x, t)$ that minimize action, we need to equate the so-called *variational derivatives* to 0:

$$\frac{\delta L}{\delta \varphi} = 0.$$

There is a general expression for energy of any such field; see, e.g., [5]. To be more precise, what is described is the so-called energy-momentum tensor T_{ij} that described the energy density. The overall energy of the field can then be determined by integrating this energy density over the whole space—just like the overall mass of a body can be obtained by integrating its density.

The formulas from [5] describe T_{ij} when the space-time is flat. It is known that in reality, the space-time is curved; see, e.g., [3, 6, 8]. In the curved space-time, there also exist formulas that describe the energy-moment tensor of each physical theory. Namely, it turns out that the energy-momentum tensor is described in terms of the variational derivatives—namely, variational derivative with respect to the metric tensor g_{ij} that describes the space-time:

$$T_{ij} = \frac{\delta L}{\delta g_{ij}}. \tag{7}$$

This formula works well for almost all physical fields, with one notable exception—it does not work for the gravitational field. Namely, if we take, as L, the Lagrangian that describes the gravitational field, then the corresponding variational equations lead to

$$\frac{\delta L}{\delta g_{ij}} = 0. \tag{8}$$

So, in view of the formula (7), we conclude that $T_{ij} = 0$—i.e., that the gravitational field carries no energy.

Of course, from the physical viewpoint, this conclusion makes no sense: as we have mentioned, the gravitational field has energy. What this conclusion actually shows is that the energy *density* T_{ij} is equal to 0. For all other fields, the overall energy can be determined as an integral of all the energy density values—in this sense, the energy is localized. In contrast, for the gravitational field, the energy cannot be described as such an integral:

- for this field, density is 0, so its integral is also 0,
- while the overall energy is non-zero.

Thus, for the gravitational field, the energy is *not* localized [6, 8].

Comment. It should be mentioned that this non-local character of gravitational energy does not depend on the theory: the same conclusion can be made if more accurate experiments will force us to replace General Relativity with some more accurate theory.

3 How Non-localness Helps to Speed Up Computations?

In the previous section, we showed that, according to some serious physicists, some physical processes are not localized. How can this non-localness help to speed up computations?

When all the processes are bounded by the speed of light c, the smallest parallel computation time is described by the right-hand side of the formula (6).

By definition, non-local processes means that some communications can spread with velocities v larger than the speed of light: $v > c$. In this case, similarly, we can derive a formula

$$t \geq \text{const} \cdot v^{-3/4} \cdot T^{1/4}. \tag{9}$$

Since $v > c$, the corresponding smallest parallel computation time—as described by the right-hand side of the formula (8)—is smaller that in the localized case. Thus, the use of non-localized physical processes can indeed speed up computations.

Acknowledgements This work was supported in part by the National Science Foundation grants 1623190 (A Model of Change for Preparing a New Generation for Professional Practice in Computer Science), and HRD-1834620 and HRD-2034030 (CAHSI Includes), and by the AT&T Fellowship in Information Technology. It was also supported by the program of the development of the Scientific-Educational Mathematical Center of Volga Federal District No. 075-02-2020-1478, and by a grant from the Hungarian National Research, Development and Innovation Office (NRDI).

References

1. J. Bell, On the Einstein Podolsky Rosen paradox. Physics **1**(3), 195–200 (1964)
2. D. Bohm, A suggested interpretation of the quantum theory in terms of "hidden variables" I. Phys. Rev. **85**(2), 166–179 (1952)
3. R. Feynman, R. Leighton, M. Sands, *The Feynman Lectures on Physics* (Addison Wesley, Boston, 2005)
4. O. Kosheleva, V. Kreinovich, Finding the best function: a way to explain calculus of variations to engineering and science students. Appl. Math. Sci. **7**(144), 7187–7192 (2013)
5. L.D. Landau, E.M. Lifshitz, *The Classical Theory of Fields* (Butterworth-Heinemann, Oxford, 1980)
6. C.W. Misner, K.S. Thorne, J.A. Wheeler, *Gravitation* (W. H. Freeman, New York, 1973)
7. D. Morgenstein, V. Kreinovich, Which algorithms are feasible and which are not depends on the geometry of space-time. Geombinatorics **4**(3), 80–97 (1995)

8. K.S. Thorne, R.D. Blandford, *Modern Classical Physics: Optics, Fluids, Plasmas, Elasticity, Relativity, and Statistical Physics* (Princeton University Press, Princeton, 2017)
9. V. Vanchurin, The world as a neural network. Entropy **22**(11), Paper 1210 (2020)
10. V. Vanchurin, Towards a theory of quantum gravity from neural networks. Entropy **24**(1), Paper 7 (2021)

General Studies of Soft Computing Techniques

Computational Paradox of Deep Learning: A Qualitative Explanation

Jonatan Contreras, Martine Ceberio, Olga Kosheleva, Vladik Kreinovich, and Nguyen Hoang Phuong

Abstract In general, the more unknowns in a problem, the more computational efforts is necessary to find all these unknowns. Interestingly, in state-of-the-art machine learning methods like deep learning, computations become easier when we increase the number of unknown parameters way beyond the number of equations. In this paper, we provide a qualitative explanation for this computational paradox.

1 Formulation of the Problem

Regression/machine learning: a brief reminder. In many practical situations, we need to find a dependence $y = f(x_1, \ldots, x_n)$ between different quantities—based on several cases $k = 1, 2, \ldots, K$ in which we know the values $x_1^{(k)}, \ldots, x_n^{(k)}$ and $y^{(k)}$ of all related quantities. In statistics, this problem is known as *regression*, in computer science, it is known as *machine learning*.

J. Contreras · M. Ceberio · V. Kreinovich (✉)
Department of Computer Science, University of Texas at El Paso, 500 W. University, El Paso, TX 79968, USA
e-mail: vladik@utep.edu

J. Contreras
e-mail: jmcontreras2@miners.utep.edu

M. Ceberio
e-mail: mceberio@utep.edu

O. Kosheleva
Department of Teacher Education, University of Texas at El Paso, 500 W. University, El Paso, TX 79968, USA
e-mail: olgak@utep.edu

N. Hoang Phuong
Artificial Intelligence Division, Information Technology Faculty, Thang Long University, Nghiem Xuan Yem Road, Hoang Mai District, Hanoi, Vietnam

N. H. Phuong and V. Kreinovich (eds.), *Deep Learning and Other Soft Computing Techniques*, Studies in Computational Intelligence 1097,
https://doi.org/10.1007/978-3-031-29447-1_22

The usual approach to solving this problem is to select a family of functions $f(x_1, \ldots, x_n, c_1, \ldots, c_m)$ and select the values of the parameters $c_1, \ldots, c_m$ for which, for all k from 1 to K, we have

$$y^{(k)} \approx f\left(x_1^{(k)}, \ldots, x_n^{(k)}, c_1, \ldots, c_m\right). \tag{1}$$

Sometimes, we select an explicit formulation of this family—e.g., in linear regression, we select linear functions:

$$f(x_1, \ldots, x_n, c_1, \ldots, c_{n+1}) = c_1 \cdot x_1 + \cdots + c_n \cdot x_n + c_{n+1}; \tag{2}$$

in other cases, we select the family of all quadratic functions, etc.

In many machine learning techniques, the corresponding family is described indirectly—e.g., in neural networks, the values c_i are weights of connections between different neurons.

It is known that the efficiency of regression/machine learning depends on the proper section of the corresponding family, in particular, on the number of parameters m.

In general, computational complexity increases with the number of unknowns. In general, the larger the number of unknowns, the more computational effort we need to solve the corresponding problem. For example, optimizing a function of one variable is easier than optimizing a function of two or three variables; solving a single equation is usually computationally easier than solving a system of several equations, etc.

Under-fitting and over-fitting: traditional statistical approach (see, e.g., [23]). As we have mentioned, the efficiency of regression/machine learning depends on the proper section of the corresponding family, in particular, on the number of parameters m.

If we select the value m to be too small, we may not get a good fit—e.g., if we only consider linear regression and the actual dependence is highly non-linear. This is known as *under-fitting*.

On the other hand, if we select the value m to be too high, e.g., if we choose $m \geq K$, then (1) becomes a system of K equations with $m \geq K$ unknowns. In general, in such cases, we can get exact equality in all K equations, i.e., we can have:

$$y^{(k)} = f\left(x_1^{(k)}, \ldots, x_n^{(k)}, c_1, \ldots, c_m\right). \tag{1a}$$

For $m > K$, we have more unknowns than equations, so we have several possible solutions for which there is exact equality (1a). The problem is that the measurement results $x_1^{(k)}, \ldots, x_n^{(k)}$ and $y^{(k)}$ come with random measurement errors, and the exact fit means that we follow this randomness—and thus, we get a not very good prediction accuracy.

For example, suppose that the actual value of y is constant $y = 1$, and we try to exactly fit the measurement results $y^{(1)} = 1.0$ and $y^{(2)} = 1.1$ corresponding to $x_1^{(1)} = 0$ and $x_1^{(2)} = 1$ with the linear dependence $y = c_1 \cdot x_1 + c_2$. Then the corresponding equations (1a), which in this case take the form

$$y^{(1)} = c_1 \cdot x^{(1)} + c_2 \text{ and } y^{(2)} = c_1 \cdot x^{(2)} + c_2$$

lead to $c_1 = 0.1$ and $c_2 = 1$, i.e., to the formula $y = 0.1 \cdot x + 1$. For large x, the predicted value can be very different from the desired value $y = 1$. This phenomenon is known as *over-fitting*.

First (accuracy-related) paradox of deep learning: going beyond supposed over-fitting increases prediction accuracy. At present, the most efficient machine learning techniques are the techniques of deep learning; see, e.g., [4]. These techniques require a large amount of data to be successful; in situations when we do not have that many available data points, other machine learning techniques work better.

Interestingly, deep learning and other state-of-the-art machine learning techniques often achieve their successes when the number of parameters m is much larger than the number K of data points—contrary to what the traditional statistics predicts. Researchers have studied this phenomenon in detail by gradually increasing m; see, e.g., [2, 3, 22]. They found out that as m increases beyond K, at first, the prediction accuracy decreases—exactly as predicted by traditional statistics. However, as the value m increases further, the prediction accuracy starts increasing again.

An explanation of the first (accuracy-related) paradox. A mathematical explanation of this paradox has been proposed in [2, 3].

Remaining problem: computational paradox of deep learning. While a solution to the accuracy-related paradox of deep learning is known, there is another deep-learning-related paradox—mentioned in [2, 3]—for which there is, at present, no satisfactory solution.

Specifically, it turned out that as the number m of parameters increases, solving the corresponding system of approximate equations—which usually means minimizing some functional (like least squares) $F(c)$, where $c \stackrel{\text{def}}{=} (c_1, \ldots, c_m)$ that describes the discrepancy between the left- and right-hand sides—becomes computationally easier; see, e.g., [21, 24].

What we do in this paper. In this paper, we provide a possible qualitative explanation for this computational paradox.

2 General Explanation

The proposed qualitative explanation is based on the following two known results.

First known result: uniqueness makes computations easier. It is known that, in general, uniqueness makes solutions easier; namely:

- for many general computational problems, there is an algorithm that solves all the cases in the which the solution is unique; see, e.g., [5–7, 9, 11, 15–17, 19]; this is true for all kinds of problems—optimization problems, solving systems of equations, etc.;
- on the other hand, for these same general classes of problems, it can be proven that no algorithm is possible that would solve all the cases in which the problem has exactly two solutions; see, e.g., [8–15, 20].

There exist good arguments that a similar phenomenon occurs for problems for which a general algorithm is possible: namely, these arguments show that, in general, cases for which there exists a unique solution are easier to solve; see, e.g., [1].

Comment. Some of these results are technically difficult, so we will not provide the proofs, we only provide links to papers where these proofs are described.

Second known result: a random function attains its optimum at a single point. The second result—which is not that technically difficult—is related to the study of random functions, i.e., reasonable probability measures on reasonable sets of functions $F(c)$. An important feature of such probability measures is that for each reasonable numerical characteristic $v(F)$ of a function $F(c)$ (such as its integral or its maximum or minimum over a certain area), this characteristic should also be "random" in the intuitive sense of this word. In particular, this means:

- that for each real number r, the probability that we have $v(F) = r$ is equal to 0, and
- that for every two different characteristics $v \neq v'$, the probability that they have the same value, i.e., that $v(F) = v'(F)$, is also equal to 0.

Let us show how this applies to minima of random functions.

Suppose that a function $F(c)$ attains its minimum value m for each two different inputs $s \neq t$:

$$F(s) = F(t) = \min_{c} F(c). \tag{3}$$

Then, we can find a hyperplane H with rational coefficients that separates these inputs s and t: namely, s is in one of the half-spaces H_- corresponding to this hyperplane, while t is in another of these two half-spaces H_+. Since $s \in H_-$ and $F(s) = m$, we have

$$m = \min_{c} F(c) \leq \min_{c \in H_-} F(c) \leq F(s) = m,$$

thus

$$\min_{c \in H_-} F(x) = m.$$

Similarly, due to $t \in H_+$ and $F(t) = m$, we have

$$\min_{c \in H_+} F(c) = m.$$

Thus, we have:

$$\min_{c \in H_-} F(c) = \min_{c \in H_+} F(c). \tag{4}$$

In line with the above-described meaning of a reasonable probability measure, the probability that two characteristics

$$\max_{c \in H_-} F(c) \text{ and } \max_{c \in H_+} F(c)$$

have equal values—as in formula (4)—is 0.

There are countably many rational numbers, so there are countably many hyperplane with rational coefficients. For each of them, the probability of equality (4) is zero. Thus, the probability that the equality (4) happens for some hyperplane with rational coefficients is also 0—since the union of countably many sets of probability measure 0 also has probability measure 0.

Thus, the probability that a function attains its minimum at two (or more) different inputs is 0. Therefore, a *random* function—i.e., a function that does not belong to any definable set of measure 0, or, alternatively, that satisfies every law that is true for almost all functions (see, e.g., [18])—cannot attain its minimum at two or more different points. So, a random function attains its minimum at a single point.

Resulting explanation. Now, we are ready to present our explanation.

In the usual statistical case, as we have mentioned earlier, when the number of parameters exceeds the number of unknowns, we have a non-unique solution to the corresponding fitting problem.

In deep learning, we do not simply make a fit, there is a lot of randomness involved: e.g., initial value of all the coefficients are random. This randomness is not just a feature of a specific training algorithm, it is inevitable: otherwise, if we start with the same deterministic values of the weights, and use some deterministic algorithm for training, all neurons in each layer will have the same weights, so their outputs will simply uselessly duplicate each other.

Because of this randomness, the actual objective function corresponding to several training steps of a neural network is random and thus, attains its minimum at only one point. So:

- in the traditional regression case, we have a problem with multiple solutions, while
- in the neural network case, we have a problem with a single solution.

Since, as we have mentioned, problems with a single solution are easier to solve than problems with multiple solutions, this explains, on the qualitative level, why optimization related to training a neural network turns out to be computationally easier than the optimization related to the usual multi-parameter regression.

3 Specific Explanation Related to Gradient-Based Training of Neural Networks

Gradient descent: a brief reminder. For neural networks, the above general explanation can be supplemented by specific details. These details are based on the fact that training of a neural network is based on gradient descent (see, e.g., [4]), when on each iteration, we replace the previous values c_i of all the parameters by a new value

$$c_i' = c_i + \Delta c_i, \tag{5}$$

where

$$\Delta c_i = -\alpha \cdot \frac{\partial F}{\partial c_i} \tag{6}$$

for some constant α. After this change, the value of the objective function changes from the previous value $F(c_1, \ldots, c_m)$ to the new value

$$F(c_1', \ldots, c_m') = F(c_1, \ldots, c_m) + \Delta F, \tag{7}$$

where we denoted:

$$\Delta F \stackrel{\text{def}}{=} F(c_1', \ldots, c_m') - F(c_1, \ldots, c_m) =$$

$$F(c_1 + \Delta c_1, \ldots, c_m + \Delta c_m) - F(c_1, \ldots, c_m). \tag{8}$$

The changes Δc_i are usually small. So, to estimate ΔF, we can expand the right-hand side of the formula (8) in Taylor series in terms of Δc_i and keep only linear terms in this expansion. As a result, we get

$$\Delta F = \sum_{i=1}^{m} \frac{\partial F}{\partial c_i} \cdot \Delta c_i. \tag{9}$$

Substituting the expression (6) into this formula, we conclude that

$$\Delta F = -\alpha \cdot \sum_{i=1}^{m} \left(\frac{\partial F}{\partial c_i}\right)^2. \tag{10}$$

Resulting explanation. As we have mentioned, to fit K equalities (1a), it is sufficient to use K parameters $c_1, \ldots, c_K$. In this case, after each iteration of the gradient descent, the value of the objective function decreases by the value

$$\Delta F_K = -\alpha \cdot \sum_{i=1}^{K} \left(\frac{\partial F}{\partial c_i} \right)^2. \tag{11}$$

If we use all $m > K$ parameters, then the resulting decrease (10) in the value of the minimized objective function $F(c)$ is even larger—and, the larger m, the larger this decrease. So, indeed, if we increase the number of coefficients, gradient descent becomes much more efficient than for smaller number of coefficients—and this is exactly what was empirically observed in [21, 24].

Acknowledgements This work was supported in part by the National Science Foundation grants 1623190 (A Model of Change for Preparing a New Generation for Professional Practice in Computer Science), and HRD-1834620 and HRD-2034030 (CAHSI Includes), and by the AT&T Fellowship in Information Technology.

It was also supported by the program of the development of the Scientific-Educational Mathematical Center of Volga Federal District No. 075-02-2020-1478, and by a grant from the Hungarian National Research, Development and Innovation Office (NRDI).

References

1. R. Beigel, H. Buhrman, L. Fortnow, NP might not be as easy as detecting unique solutions, in *Proceedings of the Thirtieth Annual ACM Symposium on Theory of Computing (STOC)*, pp. 203–208 (1998)
2. M. Belkin, Fit without fear: remarkable mathematical phenomena of deep learning through the prism of interpolation. Acta Numerica **30**, 203–248 (2021)
3. M. Belkin, D. Hsu, S. Ma, S. Mandal, Reconciling modern machine-leaning practice and the classical bias-variance trade-off. Proc. Nat. Acad. Sci. USA **116**(32), 15849–15854 (2019)
4. I. Goodfellow, Y. Bengio, A. Courville, *Deep Learning* (MIT Press, Cambridge, MA, 2016)
5. U. Kohlenbach, *Theorie der majorisierbaren und stetigen Funktionale und ihre Anwendung bei der Extraktion von Schranken aus inkonstruktiven Beweisen: Effektive Eindeutigkeitsmodule bei besten Approximationen aus ineffektiven Eindeutigkeitsbeweisen*, Ph.D. Dissertation, Frankfurt am Main (1990)
6. U. Kohlenbach, Effective moduli from ineffective uniqueness proofs. An unwinding of de La Vallée Poussin's proof for Chebycheff approximation, Annals for Pure and Applied Logic **64**(1), 27–94 (1993)
7. U. Kohlenbach, *Applied Proof Theory: Proof Interpretations and their Use in Mathematics* (Springer, Berlin-Heidelberg, 2008)
8. V. Kreinovich, *Complexity Measures: Computability and Applications*, Master thesis, Leningrad University, Department of Mathematics, Division of Mathematical Logic and Constructive Mathematics (1974) (in Russian)
9. V. Kreinovich, Uniqueness implies algorithmic computability, in *Proceedings of the 4th Student Mathematical Conference* (Leningrad University, Leningrad, 1975), pp. 19–21 (in Russian)
10. V. Kreinovich, Reviewer's remarks in a review of D. S. Bridges, *Constrictive Functional Analysis* (Pitman, London, 1979); *Zentralblatt für Mathematik*, vol. 401, pp. 22–24 (1979)

11. V. Kreinovich, *Categories of space-time models*, Ph.D. dissertation, Novosibirsk, Soviet Academy of Sciences, Siberian Branch (Institute of Mathematics, 1979) (in Russian)
12. V. Kreinovich, Unsolvability of several algorithmically solvable analytical problems. Abst. Am. Math. Soc. **1**(1), 174 (1980)
13. V. Kreinovich, *Philosophy of Optimism: Notes on the Possibility of using Algorithm Theory when describing historical processes, Leningrad Center for New Information Technology "Informatika"* (Technical report, Leningrad, 1989). ((in Russian))
14. V. Kreinovich, R.B. Kearfott, Computational complexity of optimization and nonlinear equations with interval data, in *Abstracts of the Sixteenth Symposium on Mathematical Programming with Data Perturbation* (The George Washington University, Washington, D.C., 26–27 May 1994)
15. V. Kreinovich, A. Lakeyev, J. Rohn, P. Kahl, *Computational Complexity and Feasibility of Data Processing and Interval Computations* (Kluwer, Dordrecht, 1998)
16. V. Kreinovich, K. Villaverde, Extracting computable bounds (and algorithms) from classical existence proofs: girard domains enable us to go beyond local compactness. Int. J. Intell. Technol. Appl. Stat. (IJITAS) **12**(2), 99–134 (2019)
17. D. Lacombe, Les ensembles récursivement ouvert ou fermés, et leurs applications à l'analyse récurslve. Compt Rend. **245**(13), 1040–1043 (1957)
18. M. Li, P. Vitanyi, *An Introduction to Kolmogorov Complexity and Its Applications* (Springer, Berlin, Heidelberg, New York, 2008)
19. V.A. Lifschitz, Investigation of constructive functions by the method of fillings. J. Soviet Math. **1**, 41–47 (1973)
20. L. Longpré, V. Kreinovich, W. Gasarch, G.W. Walster, m solutions good, $m-1$ solutions better. Appl. Math. Sci. **2**(5), 223–239 (2008)
21. S. Ma, R. Bassily, M. Belkin, The power of interpolation: understanding the effectiveness of SGD in modern over-parameterized learning, in *Proceedings of the 35th International Conference on Machine Learning, Proceedings of Machine Learning*, ed. by J. Dy, A. Krause, vol. 80 (Stockholmsmässan, Stockholm, Sweden, 2018), pp. 3325–3334
22. D. Monroe, A deeper understanding of deep learning: kernel methods clarify why neural networks generalize so well. Commun. ACM **65**(6), 19–20 (2022)
23. D.J. Sheskin, *Handbook of Parametric and Nonparametric Statistical Procedures* (Chapman and Hall/CRC, Boca Raton, FL, 2011)
24. M. Soltanolkotabi, A. Javanmard, J.D. Lee, Theoretical insight into the optimization landscape of over-parameterization shallow neural networks. IEEE Trans. Inf. Theory **65**, 742–769 (2018)

Graph Approach to Uncertainty Quantification

Hector A. Reyes, Cliff Joslyn, and Vladik Kreinovich

Abstract Traditional analysis of uncertainty of the result of data processing assumes that all measurement errors are independent. In reality, there may be common factor affecting these errors, so these errors may be dependent. In such cases, the independence assumption may lead to underestimation of uncertainty. In such cases, a guaranteed way to be on the safe side is to make no assumption about independence at all. In practice, however, we may have information that a few pairs of measurement errors are indeed independent—while we still have no information about all other pairs. Alternatively, we may suspect that for a few pairs of measurement errors, there may be correlation—but for all other pairs, measurement errors are independent. In both cases, unusual pairs can be naturally represented as edges of a graph. In this paper, we show how to estimate the uncertainty of the result of data processing when this graph is small.

1 Introduction

What is the problem and what we do about it: a brief description. Estimating uncertainty of the result of data processing is important in many practical applications. Corresponding formulas are well known for two extreme cases:

H. A. Reyes · V. Kreinovich (✉)
Department of Computer Science, University of Texas at El Paso, 500 W. University El Paso, El Paso, TX 79968, USA
e-mail: vladik@utep.edu

H. A. Reyes
e-mail: hareyes2@miners.utep.edu

C. Joslyn
Mathematics of Data Science, Pacific Northwest National Laboratory, 1100 Dexter Ave. N # 500, Seattle, WA 98109, USA
e-mail: cliff.joslyn@pnnl.gov

N. H. Phuong and V. Kreinovich (eds.), *Deep Learning and Other Soft Computing Techniques*, Studies in Computational Intelligence 1097,
https://doi.org/10.1007/978-3-031-29447-1_23

- when all measurement errors are independent, and
- when we have no information about the dependence.

These cases are indeed ubiquitous, but often, the actual cases are somewhat different; e.g.:

- most pairs of inputs are known to be independent, but
- there are a few pairs for which we are not sure.

Alternatively, for almost all pairs, we may have no information about the dependence, but for a few pairs of inputs, we know that the corresponding measurement errors are independent. Such unusual pairs can be naturally represented as edges of a graph. It is desirable to analyze how the presence of this graph changes the corresponding estimates.

In this paper, we start answering this question for all graphs of sizes 2, 3, and 4. We hope that our results will be extended to larger-size graphs.

Structure of the paper. In Sect. 2, we provide a detailed description of the general problem, and describe how uncertainty is estimated in the above-described two extreme cases. In the following sections, we present our results about situations in which the deviation from one of these extreme cases is described by a small-size graph.

2 Detailed Formulation of the Problem

Need for data processing. One of the main objectives of science is to describe the current state of the world and to predict future events based on what we know about the current and past states. In general, the state of a system is characterized by the values of corresponding quantities.

Some quantities we can measure directly—e.g., we can directly measure the temperature in the room or the distance between two campus buildings. However, some quantities cannot (yet) be measured directly: we cannot directly measure the temperature inside a star or a distance to this star. Since we cannot measure such quantities directly, the only way we can estimate the values of these quantities is by measuring them indirectly: i.e., by measuring related quantities $x_1, \ldots, x_n$ that are related by y by a known dependence $y = f(x_1, \ldots, x_n)$. Once we know such related quantities, we can measure their values, and use the measurement results $\widetilde{x}_1, \ldots, \widetilde{x}_n$ to compute the estimate $\widetilde{y} = f(\widetilde{x}_1, \ldots, \widetilde{x}_n)$ for y. Computing this estimate is an important case of *data processing*.

Data processing is also needed for *predictions*. For example, we may want to predict the future location of a near-Earth asteroid or tomorrow's weather. The future state can be described if we describe the future values of all the quantities characterizing this state. For example, tomorrow's weather can be characterized by temperature, wind speed, etc. To be able to make this prediction, for each of the quantities describing the future state, we need to know the relation $y = f(x_1, \ldots, x_n)$ between the future value y of this quantity and the current and past values x_i of related quantities. Once we know this relation, then we can use it to transform the measured values $\widetilde{x}_i$ of the quantities x_i into the estimate $\widetilde{y} = f(\widetilde{x}_1, \ldots, \widetilde{x}_n)$ for the desired quantity y. Computing $\widetilde{y}$ based on the measured values $\widetilde{x}_i$ is another important case of data processing.

Need for uncertainty quantification. Measurement results $\widetilde{x}$ are, in general, somewhat different from the actual (unknown) value x of the corresponding quantity; see, e.g., [5]. In other words, the difference $\Delta x \stackrel{\text{def}}{=} \widetilde{x} - x$ is usually non-zero. This difference is known as the *measurement error*.

Since the inputs $\widetilde{x}_i$ to the algorithm f are, in general, different from the actual values x_i, the resulting estimate $\widetilde{y} = f(\widetilde{x}_1, \ldots, \widetilde{x}_n)$ is, in general, different from the actual value $y = f(x_1, \ldots, x_n)$ that we would have gotten if we knew the exact values x_i. In practical situations, it is important to know how big this difference can be. For example, suppose we predict that the asteroid will pass at a distance of 150 000 km from the Earth; then:

- if the accuracy of this estimate is ±200 00 km, then this asteroid may hit the Earth, while
- if the the accuracy is ±20 000 km, this particular asteroid is harmless.

Estimating the accuracy of our estimates is an important case of *uncertainty quantification*.

What we know about measurement errors. In similar situations, with the exact same value of the measured quantity, the same measuring instrument can produce different results. This is well known to anyone who has ever repeatedly measured the same quantity: the results are always somewhat different, whether it is a current or body temperature or blood pressure. In this sense, measurement errors are *random*. Each random variable has an average (mean) value, and its actual values deviate from this mean.

Measuring instruments are usually calibrated: the measurement results provided by this instrument are compared with measurement results provided by a much more accurate ("standard") measuring instrument. If the mean difference is non-zero—i.e., in statistical terms, if the measuring instrument has a *bias*—then we can simply subtract this bias from all the measurement results and thus, make the mean error equal to 0. For example, if a person knows that his/her watch is 5 min ahead, this person can always subtract 5 min from the watch's reading and get the correct time. So, we can safely assume that the mean value $E[\Delta x]$ of each measurement error Δx is 0: $E[\Delta x] = 0$.

The deviations from the mean value are usually described by the mean squared deviation—which is known as the *standard deviation* σ. Instead of the standard deviation σ, it is sometimes convenient to use its square $V \stackrel{\text{def}}{=} \sigma^2$ which is called the *variance*. In precise terms, the variance is the mean value of the square of the difference between the random variable and its mean value: $V[X] = E\left[(X - E[X])^2\right]$. For measurement error, the mean is $E[\Delta x] = 0$, so we get a simplified formula $V[\Delta x] = E[(\Delta x)^2]$.

For each measuring instrument, the standard deviation is also determined during the calibration. So, we can assume that for each measuring instrument:

- we know that the mean value of its measurement error is 0, and
- we know the standard deviation of the measurement error.

In many cases, distributions are normal. In most practical cases, there are many factors that contribute to the measurement error. For example, if we measure voltage, the measuring instrument is affected not only by the current that we measure, but also by the currents of multiple devices present in the room, including the computer used to process the data, the lamps in the ceiling, etc. Each of these factors may be relatively small, but there are many of them, and thus, the resulting measurement error is much larger than each of them.

It is known—see, e.g., [6]—that the probability distribution of the joint effect of a large number of small random factors is close to Gaussian (normal). Thus, in such cases, we can safely assume that the measurement errors are normally distributed.

Possibility of linearization. In general, the estimation error is equal to $\Delta y \stackrel{\text{def}}{=} \widetilde{y} - y$. Here, $\widetilde{y} = f(\widetilde{x}_1, \ldots, \widetilde{x}_n)$ and $y = f(x_1, \ldots, x_n)$, so

$$\Delta y = f(\widetilde{x}_1, \ldots, \widetilde{x}_n) - f(x_1, \ldots, x_n),$$

By definition of the measurement error Δx_i as the difference $\Delta x_i = \widetilde{x}_i - x_i$, we have $x_i = \widetilde{x}_i - \Delta x_i$. Thus, the above expression for the approximation error takes the form

$$\Delta y = f(\widetilde{x}_1, \ldots, \widetilde{x}_n) - f(\widetilde{x}_1 - \Delta x_1, \ldots, \widetilde{x}_n - \Delta x_n). \tag{1}$$

Measurement errors are usually relatively small, so that terms quadratic in these errors can be safely ignored. For example, if the measurement error is 10%, its square is 1%, which is much smaller. Thus, we can expand the right-hand side of the equality (1) in Taylor series and keep only linear terms in this expansion. As a result, we conclude that

$$\Delta y = \sum_{i=1}^{n} c_i \cdot \Delta x_i, \tag{2}$$

where we denoted

$$c_i \stackrel{\text{def}}{=} \frac{\partial f}{\partial x_i}_{|x_1=\widetilde{x}_1,\ldots,x_n=\widetilde{x}_n}.$$

In other words, the desired estimation error Δy is a linear combination of measurement errors Δx_i.

Case when all measurement errors are independent. It is known that the variance of the sum of the several random variables is equal to the sum of their variances. It is also known that if we multiply a random variable by a constant, then its standard deviation is multiplied by the absolute value of this constant. So, if we denote the standard deviation of the i-th measuring instrument by σ_i, then the standard deviation of the product $c_i \cdot \Delta x_i$ is equal to $|c_i| \cdot \sigma_i$ and thus, its variance is equal to $(|c_i| \cdot \sigma_i)^2 = c_i^2 \cdot \sigma_i^2$. Thus, the variance of the sum Δy is equal to the sum of these variances:

$$\sigma^2 = \sum_{i=1}^{n} c_i^2 \cdot \sigma_i^2, \tag{3}$$

and thus, the standard deviation is equal to

$$\sigma = \sqrt{\sum_{i=1}^{n} c_i^2 \cdot \sigma_i^2}. \tag{4}$$

Towards the general case: a known geometric interpretation of random variables. We have n random variables $v_i \stackrel{\text{def}}{=} c_i \cdot \Delta x_i$. For each variable, we know its standard deviation $|c_i| \cdot \sigma_i$, and we are interested in estimating the standard deviation of the sum $\Delta y = v_1 + \cdots + v_n$ of these variables.

It is known (see, e.g., [6]) that we can interpret each variable—and, correspondingly, each linear combination of the variables—as vectors $\mathbf{a}$, $\mathbf{b}$ in an n-dimensional space, so that the length $\|\mathbf{a}\| = \sqrt{\mathbf{a} \cdot \mathbf{a}}$ of each vector (where $\mathbf{a} \cdot \mathbf{b}$ means dot (scalar) product) is equal to the standard deviation of the corresponding variable. In these terms, independence means that the two vectors are orthogonal. Indeed:

- In statistical terms, independence implies that the variance of the sum if equal to the sum of the variances.
- For the sum $\mathbf{a} + \mathbf{b}$ of two vectors, the square of the length has the form

  $$\|\mathbf{a} + \mathbf{b}\|^2 = (\mathbf{a} + \mathbf{b}) \cdot (\mathbf{a} + \mathbf{b}) = \mathbf{a} \cdot \mathbf{a} + \mathbf{b} \cdot \mathbf{b} + 2\mathbf{a} \cdot \mathbf{b}.$$

  Here, $\mathbf{a} \cdot \mathbf{a} = \|\mathbf{a}\|^2$, $\mathbf{b} \cdot \mathbf{b} = \|\mathbf{b}\|^2$, and $\mathbf{a} \cdot \mathbf{b} = \|\mathbf{a}\| \cdot \|\mathbf{b}\| \cdot \cos(\theta)$, where θ is the angle between the two vectors, so the above expression takes the form

  $$\|\mathbf{a} + \mathbf{b}\|^2 = \|\mathbf{a}\|^2 + \|\mathbf{b}\|^2 + 2\|\mathbf{a}\| \cdot \|\mathbf{b}\| \cdot \cos(\theta).$$

  So, the variance $\|\mathbf{a} + \mathbf{b}\|^2$ of the sum is equal to the sum $\|\mathbf{a}\|^2 + \|\mathbf{b}\|^2$ of the variances if and only if $\cos(\theta) = 0$, i.e., if only if the angle is 90°, and the vectors are orthogonal.

In the independent case, n vectors $\mathbf{v}_i$ corresponding to individual measurements are orthogonal to each other, so, similarly to the above argument, one can show that the length of their sum is equal to the square root of the sum of the squares of their lengths:

$$\|\mathbf{v}_1 + \cdots + \mathbf{v}_n\|^2 = \|\mathbf{v}_1\|^2 + \cdots + \|\mathbf{v}_n\|^2.$$

Let us use this geometric interpretation to estimate the uncertainty in situations when we know nothing about correlation between different measurement errors.

What if we have no information about correlations: analysis of the problem and the resulting formula. In this case, we still have n vectors $\mathbf{v}_1, \ldots, \mathbf{v}_n$ of given lengths $\|\mathbf{v}_i\| = |c_i| \cdot \sigma_i$. The main difference from the independent case is that these vectors are not necessarily orthogonal, we can have different angles between them. In this case, in contrast to the independent case, the length of the sum is not uniquely determined. For example:

- if two vectors of equal length are parallel, the length of their sum is double the length of each vector, but
- if they are anti-parallel $\mathbf{b} = -\mathbf{a}$, then their sum has length 0.

In such cases, it is reasonable to find the worst possible standard deviation, i.e., the largest possible length.

One can easily check that the sum of several vectors of given length is the largest when all these vectors are parallel and oriented in the same direction. In this case, the length of the sum is simply equal to the sum of the lengths, so we get

$$\sigma = \sum_{i=1}^{n} |c_i| \cdot \sigma_i, \tag{5}$$

and

$$\sigma^2 = \left(\sum_{i=1}^{n} |c_i| \cdot \sigma_i \right)^2. \tag{6}$$

Precise mathematical formulation of this result. The above result can be presented in the following precise form.

Definition 1

- Let $s = (\sigma_1, \ldots, \sigma_n)$ be a tuple of non-negative real numbers.
- Let D denote the class of all possible multi-D distributions $(\Delta x_1, \ldots, \Delta x_n)$ for which, for each i, we have $E[\Delta x_i] = 0$ and $\sigma[\Delta x_i] = \sigma_i$.
- Let $\mathcal{S}$ be a subset of the set D; we will denote it, as usual, by $\mathcal{S} \subseteq D$.
- Let $c = (c_1, \ldots, c_n)$ be a tuple of real numbers.
- For each distribution from D, let Δy denote $\Delta y \stackrel{\text{def}}{=} c_1 \cdot \Delta x_1 + \cdots + c_n \cdot \Delta x_n$.

Then, by $\sigma\,(c, s, \mathcal{S})$ we denote the largest possible value of the standard deviation $\sigma_\rho[\Delta y]$ over all distributions from the set $\mathcal{S}$:

$$\sigma\,(s, \mathcal{S}, c) \stackrel{\text{def}}{=} \max_{\rho \in \mathcal{S}} \sigma_\rho[\Delta y].$$

In these terms, the above result takes the following form:

Proposition 1 *For the set $\mathscr{S} = D$ of all possible distributions, we have*

$$\sigma\,(s, D, c) = \sum_{i=1}^{n} |c_i| \cdot \sigma_i.$$

Similarly, the previous result—about independent case—takes the following form.

Definition 2 By $I \subset D$, we denote the class of all distributions for which, for all i and j, the variables Δx_i and Δx_j are independent. We will call I *independent set.*

Proposition 2 *For the independent set I, we have*

$$\sigma\,(s, I, c) = \sqrt{\sum\nolimits_{i=1}^{n} c_i^2 \cdot \sigma_i^2}.$$

Comment. Interestingly, the formula (5) is similar to what we get in the linearized version of the interval case (see, e.g., [1–5]), i.e., the case when we only know the upper bound Δx_i on the absolute value of the measurement error. In other words, this means that:

- the measurement error is located somewhere in the interval $[-\Delta x_i, \Delta x_i]$, and
- we have no information about the probability of different values from this interval.

In this case, the largest possible value of the estimation error

$$\Delta y = c_1 \cdot \Delta x_1 + \cdots + c_n \cdot \Delta x_n$$

is equal to $|c_1| \cdot \Delta_1 + \cdots + |c_n| \cdot \Delta_n$. This is indeed the same expression as our formula (5).

3 What if a Few Pairs of Measurement Errors are Not Necessarily Independent

3.1 Description of the Situation

Descriptions of the situation. In the previous text, we considered two extreme cases:

- when we know that all measurement errors are independent, and
- when we have no information about their dependence.

Such cases are indeed frequent, but sometimes, situations are similar but not exactly the same. For example, we can have the case of "almost independence", when for most pairs, we know that they are independent, but for a few pairs, we do not have this information. This is the situation that we describe in this section.

Comment. The opposite situations, when we only have independence information about a few pairs, is described in the next section.

Graph representation of such situations. To describe such situations, we need to know for which pairs of measurement errors, we do not have information about independence. A natural way to represent such information is by an undirected graph in which:

- measurement errors are vertices and
- an edge connects pairs for which we do not have information about independence.

We only need to know which vertices are connected. So, it makes sense to include, in the description of the graph, only vertices that are connected by some edge, i.e., only measurement errors that may not be independent with respect to others. In this case, we arrive at the following definition.

Definition 3

- Let $G = (V, E)$ be an undirected graph with the set of vertices $V \subseteq \{1, \ldots, n\}$ for which every vertex is connected to some other vertex. Here, E is a subset $E \subseteq V \times V$ for which:
  - for each $a \in V$, we have $(a, a) \notin E$,
  - for each a and b, $(a, b) \in E$ if and only if $(b, a) \in E$, and
  - for each $a \in V$, we have $(a, b) \in E$ for some $b \in V$.
- By $I_G \subseteq D$, we mean the class of all distributions for which for all pairs $(i, j) \notin E$ the variables Δx_i and Δx_j are independent.

Discussion. Our objective is to find the value $\sigma(s, I_G, c)$ for different graphs G. In this paper, we only consider the simplest graphs: all graphs with 2, 3, or 4 vertices. We hope that this work will be extended to larger-size graphs.

3.2 General Results

Let us first present some general results. For this purpose, let us introduce the following notations. For any set $S \subseteq \{1, \ldots, n\}$, by a restriction c_S we mean sub-tuples consisting only of elements c_i for which $i \in S$. For example, for $c = (c_1, c_2, c_3, c_4)$ and $S = \{1, 3\}$, we have $c_S = (c_1, c_3)$. Similarly, we can define the restriction s_S. It is then relatively easy to show that the following result holds:

Proposition 3 *For each graph $G = (V, E)$, we have:*

$$\sigma^2 (s, I_G, c) = \sum_{i \notin V} c_i^2 \cdot \sigma_i^2 + (\sigma (s_V, I_G, c_V))^2 .$$

Comments.

- In other words, it is sufficient to only consider measurement errors from the exception set V—which are not necessarily independent, then all other measurement errors can be treated the same way as in the case when all measurement errors are independent.
- For reader's conveniences, all the proofs are placed in a special Proofs section.

Another easy-to-analyze important case is when the graph G is disconnected, i.e., consists of several connected components.

Proposition 4 *When the graph $G = (V, E)$ consists of several connected components $G_1 = (V_1, E_1), \ldots, G_k = (V_k, E_k)$ with for which $V = V_1 \cup \ldots \cup V_k$, then*

$$\sigma^2 (s, I_G, c) = \sum_{i \notin V} c_i^2 \cdot \sigma_i^2 + \sum_{j=1}^{k} \left(\sigma \left(s_{V_j}, \mathscr{S}_{G_j}, c_{V_j}\right)\right)^2 .$$

Comment. In view of this result, it is sufficient to estimate the value $\sigma (s, I_G, c)$ for connected graphs. We consider connected graphs with 2, 3, or 4 vertices.

3.3 Connected Graph of Size 2

There is only one connected graph of size 2: two vertices i_1 and i_2 connected by an edge, so that $V = \{i_1, i_2\}$ and $E = \{(i_1, i_2), (i_2, i_1)\}$.

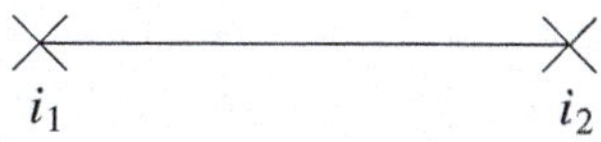

Proposition 5 *When the graph $G = (V, E)$ consists of two vertices i_1 and i_2 connected by an edge, then*

$$\sigma^2 (s, I_G, c) = \sum_{i \neq i_1, i_2} c_i^2 \cdot \sigma_i^2 + \left(|c_{i_1}| \cdot \sigma_{i_1} + |c_{i_2}| \cdot \sigma_{i_2}\right)^2 .$$

3.4 Connected Graphs of Size 3

In a connected graph of size 3, two vertices are connected, and the third vertex is:

- either connected to both of them—in this case we have a complete 3-element graph,

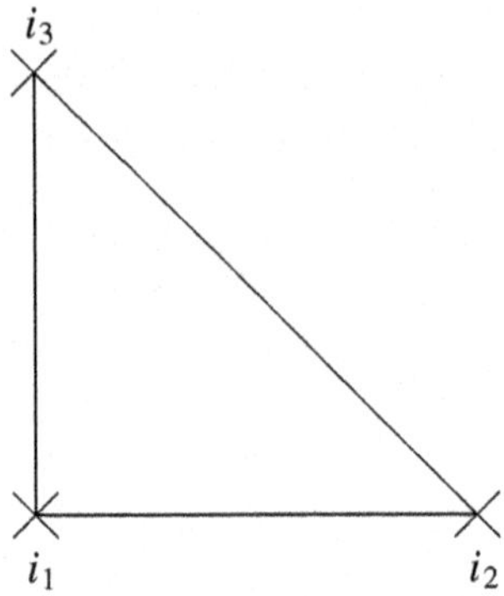

- or to only one of them.

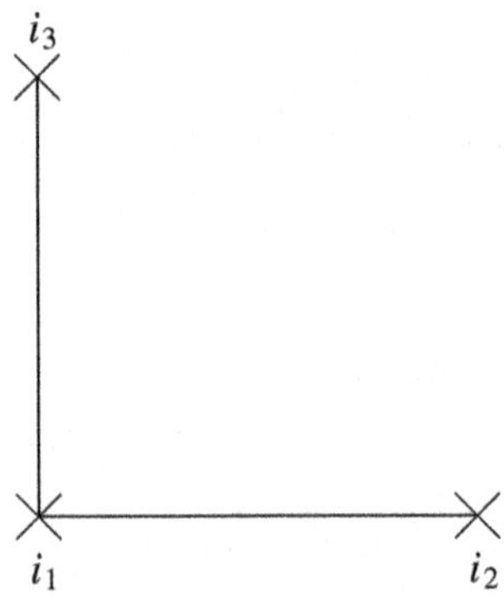

So, modulo isomorphism, there are two different connected graphs of size 3. For these graphs, we get the following results:

Proposition 6 *When* $G = (V, E)$ *is a complete 3-element graph with* $V = \{i_1, i_2, i_3\}$, *then*

$$\sigma^2(s, I_G, c) = \sum_{i \notin V} c_i^2 \cdot \sigma_i^2 + \left(|c_{i_1}| \cdot \sigma_{i_1} + |c_{i_2}| \cdot \sigma_{i_2} + |c_{i_3}| \cdot \sigma_{i_3}\right)^2.$$

Proposition 7 *For a 3-element graph with $V = \{i_1, i_2, i_3\}$ in which i_1 is connected to i_2 and i_3 but i_2 and i_3 are not connected, we have:*

$$\sigma^2(s, I_G, c) = \sum_{i \notin V} c_i^2 \cdot \sigma_i^2 + \left(|c_{i_1}| \cdot \sigma_{i_1} + \sqrt{c_{i_2}^2 \cdot \sigma_{i_2}^2 + c_{i_3}^2 \cdot \sigma_{i_3}^2} \right)^2 .$$

3.5 Connected Graphs of Size 4

Let us first consider graphs of size 4 for which there is a vertex (we will denote it i_1) connected with all three other vertices. In this case, there are four possible options:

- when connections between i_1 and all three other vertices are the only connections:

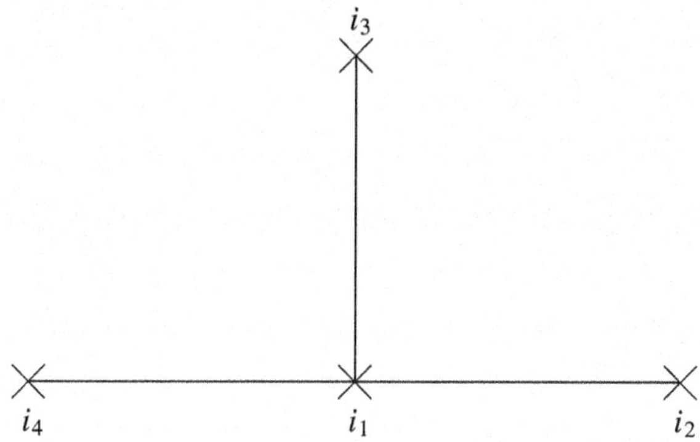

- when, in addition to edges connecting i_1 to three other vertices, there is also one connection between two of these other vertices:

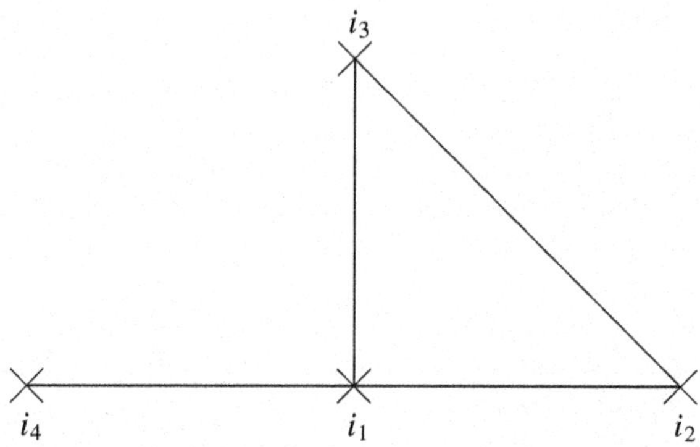

- when, in addition to edges connecting i_1 to three other vertices, there are two connection between these other vertices:

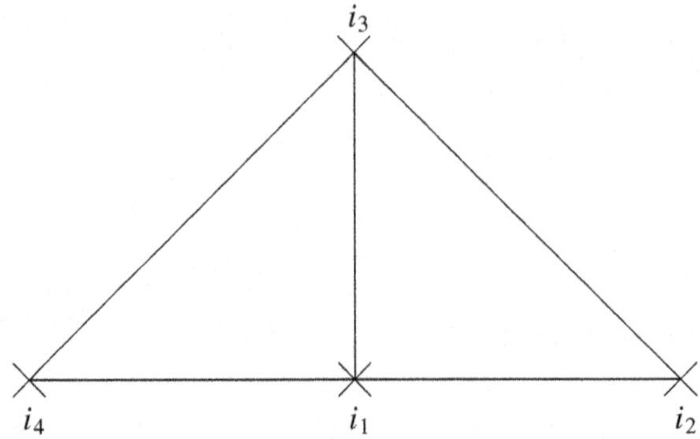

- and when we have a complete 4-element graph:

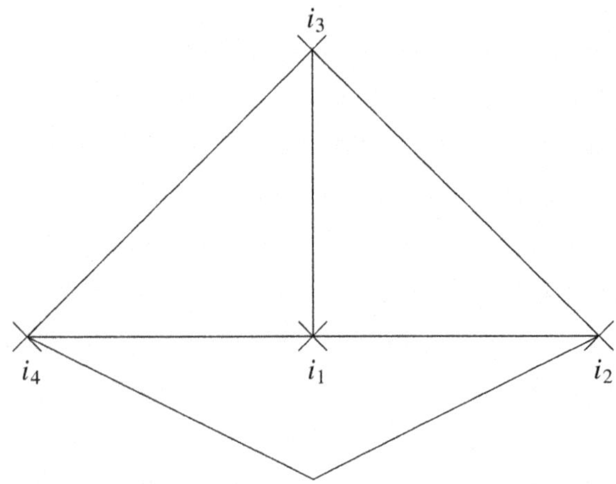

Finally, let us consider graphs in which each vertex is connected to no more than two others. If each vertex is connected to only one vertex, then a vertex i_1 is connected to some vertex i_2, and there is no space for each of them to have any other connection—so the 4-element graph containing vertices i_1 and i_2 cannot be connected. Thus, there should be at least one vertex connected to two others.

Let us denote one such vertex by i_2, and the two vertices to which i_2 is connected by i_1 and i_3. Since the graph is connected, the fourth vertex i_4 must be connected to one of the previous three vertices. The vertex i_4 cannot be connected to i_2—because then i_2 should be connected to three other vertices, and we consider the case when each vertex is connected to no more than two others. Thus, i_4 is connected to i_1 and/or i_3. If it is connected to i_1, then we can swap the names of vertices i_1 and i_3, and get the same configuration as when i_4 is connected to i_3. If i_4 is connected to both i_1 and i_3, then the resulting graph is uniquely determined. Thus, under the assumption that each vertex is connected to no more than two others, we have two possible graphs:

- a "linear" graph:

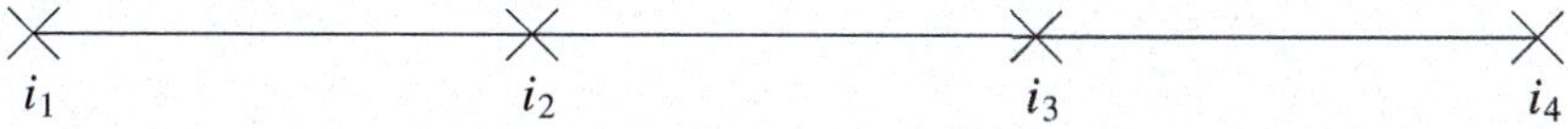

- and a "square graph":

For all these graphs, we have the following results:

Proposition 8 *For a 4-element graph with $V = \{i_1, i_2, i_3, i_4\}$, in which i_1 is connected to i_2, i_3, and i_4, but i_2, i_3, and i_4 are not connected to each other, we have:*

$$\sigma^2(s, I_G, c) = \sum_{i \notin V} c_i^2 \cdot \sigma_i^2 + \left(|c_{i_1}| \cdot \sigma_{i_1} + \sqrt{c_{i_2}^2 \cdot \sigma_{i_2}^2 + c_{i_3}^2 \cdot \sigma_{i_3}^2 + c_{i_4}^2 \cdot \sigma_{i_4}^2} \right)^2.$$

Proposition 9 *For a 4-element graph with $V = \{i_1, i_2, i_3, i_4\}$, in which i_1, i_2, and i_3 form a complete graph, and i_4 is connected only to i_1, we have:*

$$\sigma^2(s, I_G, c) = \sum_{i \notin V} c_i^2 \cdot \sigma_i^2 + \left(|c_{i_1}| \cdot \sigma_{i_1} + \sqrt{\left(|c_{i_2}| \cdot \sigma_{i_2} + |c_{i_3}| \cdot \sigma_{i_3}\right)^2 + c_{i_4}^2 \cdot \sigma_{i_4}^2} \right)^2.$$

Proposition 10 *For a 4-element graph with $V = \{i_1, i_2, i_3, i_4\}$, in which i_1, i_2, and i_3 form a complete graph, and i_4 is corrected to i_1 and i_3, we have:*

$$\sigma^2(s, I_G, c) = \sum_{i \notin V} c_i^2 \cdot \sigma_i^2 + \left(|c_{i_1}| \cdot \sigma_{i_1} + |c_{i_3}| \cdot \sigma_{i_3} + \sqrt{c_{i_2}^2 \cdot \sigma_{i_2}^2 + c_{i_4}^2 \cdot \sigma_{i_4}^2} \right)^2.$$

Proposition 11 *For a complete 4-element graph with $V = \{i_1, i_2, i_3, i_4\}$, we have:*

$$\sigma^2(s, I_G, c) = \sum_{i \notin V} c_i^2 \cdot \sigma_i^2 + \left(|c_{i_1}| \cdot \sigma_{i_1} + |c_{i_2}| \cdot \sigma_{i_2} + |c_{i_3}| \cdot \sigma_{i_3} + |c_{i_4}| \cdot \sigma_{i_4}\right)^2.$$

Proposition 12 *For a linear 4-element graph with $V = \{i_1, i_2, i_3, i_4\}$, the value $\sigma(s, I_G, c)$ has the following form:*

- *if $|c_{i_2}| \cdot \sigma_{i_2} \cdot |c_{i_3}| \cdot \sigma_{i_3} > |c_{i_1}| \cdot \sigma_{i_1} \cdot |c_{i_4}| \cdot \sigma_{i_4}$, then*

$$\sigma^2(s, I_G, c) = \sum_{i \notin V} c_i^2 \cdot \sigma_i^2 + \left(\sqrt{c_{i_1}^2 \cdot \sigma_{i_1}^2 + c_{i_2}^2 \cdot \sigma_{i_2}^2} + \sqrt{c_{i_3}^2 \cdot \sigma_{i_3}^2 + c_{i_4}^2 \cdot \sigma_{i_4}^2}\right)^2;$$

- *otherwise, we get*

$$\sigma^2(s, I_G, c) = \sum_{i \notin V} c_i^2 \cdot \sigma_i^2 + \max(V_2, V_3, V_0), \text{ where}$$

$$V_2 = c_{i_3}^2 \cdot \sigma_{i_3}^2 + \left(\sqrt{c_{i_1}^2 \cdot \sigma_{i_1}^2 + c_{i_2}^2 \cdot \sigma_{i_2}^2} + |c_{i_4}| \cdot \sigma_{i_4}\right)^2,$$

$$V_3 = c_{i_2}^2 \cdot \sigma_{i_2}^2 + \left(|c_{i_1}| \cdot \sigma_{i_1} + \sqrt{c_{i_3}^2 \cdot \sigma_{i_3}^2 + c_{i_4}^2 \cdot \sigma_{i_4}^2}\right)^2, \text{ and}$$

$$V_0 = \left(|c_{i_1}| \cdot \sigma_{i_1} + |c_{i_3}| \cdot \sigma_{i_3}\right)^2 + \left(|c_{i_2}| \cdot \sigma_{i_2} + |c_{i_4}| \cdot \sigma_{i_4}\right)^2.$$

Proposition 13 *For a square 4-element graph with $V = \{i_1, i_2, i_3, i_4\}$, we have:*

$$\sigma^2(s, I_G, c) = \sum_{i \notin V} c_i^2 \cdot \sigma_i^2 +$$

$$\left(\sqrt{c_{i_1}^2 \cdot \sigma_{i_1}^2 + c_{i_3}^2 \cdot \sigma_{i_3}^2} + \sqrt{c_{i_2}^2 \cdot \sigma_{i_2}^2 + c_{i_4}^2 \cdot \sigma_{i_4}^2}\right)^2.$$

4 What if only a Few Pairs of Measurement Errors are Known to be Independent

4.1 Description of the Situation

Graph representation of such situations. To describe such situations, we need to know for which pairs of measurement errors, we have information about independence. A natural way to represent such information is by an undirected graph in which:

- measurement errors are vertices and
- an edge connects pairs for which we have information about independence.

For simplicity, we can only consider vertices that are connected by some edge, i.e., only measurement errors that are known to be independent with respect to others. In this case, we arrive at the following definition.

Definition 4

- Let $G = (V, E)$ be an undirected graph with the set of vertices $V \subseteq \{1, \ldots, n\}$. Here, E is a subset $E \subseteq V \times V$ for which:
  - for each $a \in V$, we have $(a, a) \notin E$, and
  - for each a and b, $(a, b) \in E$ if and only if $(b, a) \in E$.
- By $D_G \subseteq D$, we mean the class of all distributions for which for all pairs $(i, j) \in E$ the variables Δx_i and Δx_j are independent.

Discussion. Our objective is to find the value $\sigma(s, D_G, c)$ for different graphs G. In this paper, we only consider the simplest graphs: all graphs with 2, 3, or 4 vertices. We hope that this work will be extended to larger-size graphs.

4.2 General Results

Let us first present some general results. It is then relatively easy to show that the following result holds:

Proposition 14 *For each graph $G = (V, E)$, we have:*

$$\sigma(s, D_G, c) = \sum_{i \notin V} |c_i| \cdot \sigma_i + \sigma(s_V, D_G, c_V).$$

Comment. In other words, it is sufficient to only consider measurement errors from the exception set V—which are not necessarily independent; then all other measurement errors can be treated the same way as in the case when we have no information about dependence.

Another easy-to-analyze important case is when the graph G is disconnected, consisting of several connected components.

Proposition 15 *When the graph $G = (V, E)$ consists of several connected components $G_1 = (V_1, E_1), \ldots, G_k = (V_k, E_k)$ with for which $V = V_1 \cup \ldots \cup V_k$, then*

$$\sigma(s, D_G, c) = \sum_{i \notin V} |c_i| \cdot \sigma_i + \sum_{j=1}^{k} \sigma\left(s_{V_j}, \mathscr{S}_{G_j}, c_{V_j}\right).$$

Comment. In view of this result, it is sufficient to estimate the value $\sigma(s, D_G, c)$ for connected graphs. In this paper, we consider all connected graphs with 2, 3, or 4 vertices.

4.3 Connected Graph of Size 2

Proposition 16 *When the graph $G = (V, E)$ consists of two vertices i_1 and i_2 connected by an edge, then*

$$\sigma^2(s, D_G, c) = \sum_{i \notin V} |c_i| \cdot \sigma_i + \sqrt{c_{i_1}^2 \cdot \sigma_{i_1}^2 + c_{i_2}^2 \cdot \sigma_{i_2}^2}.$$

4.4 Connected Graphs of Size 3

Proposition 17 *When $G = (V, E)$ is a complete 3-element graph with $V = \{i_1, i_2, i_3\}$, then*

$$\sigma^2(s, D_G, c) = \sum_{i \notin V} |c_i| \cdot \sigma_i + \sqrt{c_{i_1}^2 \cdot \sigma_{i_1}^2 + c_{i_2}^2 \cdot \sigma_{i_2}^2 + c_{i_3}^2 \cdot \sigma_{i_3}^2}.$$

Proposition 18 *For a 3-element graph with $V = \{i_1, i_2, i_3\}$, in which i_1 is connected to i_2 and i_3 but i_2 and i_3 are not connected, we have:*

$$\sigma^2(s, D_G, c) = \sum_{i \notin V} |c_i| \cdot \sigma_i + \sqrt{c_{i_1}^2 \cdot \sigma_{i_1}^2 + \left(|c_{i_2}| \cdot \sigma_{i_2} + |c_{i_3}| \cdot \sigma_{i_3}\right)^2}.$$

4.5 Connected Graphs of Size 4

Proposition 19 *For a 4-element graph with $V = \{i_1, i_2, i_3, i_4\}$, in which i_1 is connected to i_2, i_3, and i_4, but i_2, i_3, and i_4 are not connected to each other, we have:*

$$\sigma(s, D_G, c) = \sum_{i \notin V} |c_i| \cdot \sigma_i + \sqrt{c_{i_1}^2 \cdot \sigma_{i_1}^2 + \left(|c_{i_2}| \cdot \sigma_{i_2} + |c_{i_3}| \cdot \sigma_{i_3} + |c_{i_4}| \cdot \sigma_{i_4}\right)^2}.$$

Proposition 20 *For a 4-element graph with $V = \{i_1, i_2, i_3, i_4\}$, in which i_1, i_2, and i_3 form a complete graph, and i_4 is connected only to i_1, we have:*

$$\sigma(s, D_G, c) = \sum_{i \notin V} |c_i| \cdot \sigma_i + \sqrt{c_{i_1}^2 \cdot \sigma_{i_1}^2 + \left(\sqrt{c_{i_2}^2 \cdot \sigma_{i_2}^2 + c_{i_3}^2 \cdot \sigma_{i_3}^2} + |c_{i_4}| \cdot \sigma_{i_4}\right)^2}.$$

Proposition 21 *For a 4-element graph with $V = \{i_1, i_2, i_3, i_4\}$, in which i_1, i_2, and i_3 form a complete graph, and i_4 is corrected to i_1 and i_3, we have:*

$$\sigma(s, D_G, c) = \sum_{i \notin V} |c_i| \cdot \sigma_i + \sqrt{c_{i_1}^2 \cdot \sigma_{i_1}^2 + c_{i_3}^2 \cdot \sigma_{i_3}^2 + \left(|c_{i_2}| \cdot \sigma_{i_2} + |c_{i_4}| \cdot \sigma_{i_4}\right)^2}.$$

Proposition 22 *For a complete 4-element graph with $V = \{i_1, i_2, i_3, i_4\}$, we have:*

$$\sigma(s, D_G, c) = \sum_{i \notin V} |c_i| \cdot \sigma_i + \sqrt{c_{i_1}^2 \cdot \sigma_{i_1}^2 + c_{i_2}^2 \cdot \sigma_{i_2}^2 + c_{i_3}^2 \cdot \sigma_{i_3}^2 + c_{i_4}^2 \cdot \sigma_{i_4}^2}.$$

Proposition 23 *For a linear 4-element graph with $V = \{i_1, i_2, i_3, i_4\}$:*

- *if $|c_{i_1}| \cdot \sigma_{i_1} \cdot |c_{i_4}| \cdot \sigma_{i_4} > |c_{i_2}| \cdot \sigma_{i_2} \cdot |c_{i_3}| \cdot \sigma_{i_3}$, then*

$$\sigma(s, D_G, c) = \sum_{i \notin V} |c_i| \cdot \sigma_i + \sqrt{c_{i_1}^2 \cdot \sigma_{i_1}^2 + c_{i_2}^2 \cdot \sigma_{i_2}^2} + \sqrt{c_{i_3}^2 \cdot \sigma_{i_3}^2 + c_{i_4}^2 \cdot \sigma_{i_4}^2}.$$

- *otherwise, we get*

$$\sigma(s, D_G, c) = \sum_{i \notin V} |c_i| \cdot \sigma_i + \sqrt{\max(\sigma_2^2, \sigma_3^2, \sigma_0^2)}, \text{ where}$$

$$\sigma_2^2 = v_{i_3}^2 + \left(\sqrt{v_{i_1}^2 + v_{i_2}^2} + v_{i_4}\right)^2,$$

$$\sigma_3^2 = v_{i_2}^2 + \left(v_{i_1} + \sqrt{v_{i_3}^2 + v_{i_4}^2}\right)^2, \textit{ and}$$

$$\sigma_0^2 = \left(v_{i_1} + v_{i_3}\right)^2 + \left(v_{i_2} + v_{i_4}\right)^2.$$

Proposition 24 *For a square 4-element graph with* $V = \{i_1, i_2, i_3, i_4\}$, *we have:*

$$\sigma\left(s, D_G, c\right) = \sum_{i \notin V} |c_i| \cdot \sigma_i +$$

$$\sqrt{\left(|c_{i_1}| \cdot \sigma_{i_1} + |c_{i_3}| \cdot \sigma_{i_3}\right)^2 + \left(|c_{i_2}| \cdot \sigma_{i_2} + |c_{i_4}| \cdot \sigma_{i_4}\right)^2}.$$

Hypothesis. On all these cases, the desired bounds can be obtained if we combine the values $|c_i| \cdot \sigma_i$ by using either formulas corresponding to independence, or formulas corresponding to the possibility of dependence. For example, in the case covered by Proposition 24:

- we first combine $|c_{i_1}| \cdot \sigma_{i_1}$ and $|c_{i_3}| \cdot \sigma_{i_3}$ by using the formula corresponding to the possibility of dependence,
- then, we combine $|c_{i_1}| \cdot \sigma_{i_1}$ and $|c_{i_3}| \cdot \sigma_{i_3}$ by using the formula corresponding to the possibility of dependence,
- and finally, we combine the two previous results by using the formula corresponding to independence.

In some cases—e.g., in the case covered by Proposition 12—we have subcases in which different formulas of this type should be used. In such situations, the selection of these subcases can be also described by inequalities relating such formulas. Indeed, in the case of Proposition 12, the inequality

$$|c_{i_2}| \cdot \sigma_{i_2} \cdot |c_{i_3}| \cdot \sigma_{i_3} > |c_{i_1}| \cdot \sigma_{i_1} \cdot |c_{i_4}| \cdot \sigma_{i_4}$$

describing which subcase to use is equivalent to

$$(|c_{i_2}| \cdot \sigma_{i_2} + |c_{i_3}| \cdot \sigma_{i_3})^2 + c_{i_1}^2 \cdot \sigma_{i_1}^2 + c_{i_4}^2 \cdot \sigma_{i_4}^2 >$$

$$c_{i_2}^2 \cdot \sigma_{i_2}^2 + c_{i_3}^2 \cdot \sigma_{i_3}^2 + (|c_{i_1}^2| \cdot \sigma_{i_1} + |c_{i_4}| \cdot \sigma_{i_4})^2.$$

It is natural to conjecture that formulas corresponding to general graphs will have similar structure.

5 Proofs

Proof of Proposition 3. The proof of this proposition naturally follows from the geometric interpretation, in which we associate a vector to each random variable, and we are looking for a configuration in which the sum of these vectors has the largest length. Here, the sum $\mathbf{v}$ of all the corresponding vectors $\mathbf{v} = \mathbf{v}_1 + \cdots + \mathbf{v}_n$ can be represented as

$$\sum_{i \notin V} \mathbf{v}_i + \mathbf{a},$$

where

$$\mathbf{a} \stackrel{\text{def}}{=} \sum_{j \in V} \mathbf{v}_j.$$

Vectors $\mathbf{v}_i$ corresponding to "normal" errors ($i \notin V$) are orthogonal (since the corresponding measurement errors are independent), and since each of them is orthogonal to each of the "abnormal" vectors $\mathbf{v}_j$, it is also orthogonal to the sum $\mathbf{a}$ of these abnormal vectors. Thus, the square of the length of the sum $\mathbf{v}$ is equal to the sum of the squares of the lengths of the "normal" vectors $\mathbf{v}_i$ and of the vector $\mathbf{a}$:

$$\|\mathbf{v}\|^2 = \sum_{i \notin V} \|\mathbf{v}_i\|^2 + \|\mathbf{a}\|^2.$$

The values $\|\mathbf{v}_i\|^2$ are given: they are equal to $c_i^2 \cdot \sigma_i^2$ Thus, the largest possible value of $\|\mathbf{v}\|$ is attained when the length $\|\mathbf{a}\|$ is the largest. This largest length is what in Definitions 1 and 3 we denoted by $\sigma\,(s_V, I_G, c_V)$. Thus, we get the desired formula.

The proposition is proven.

Proof of Proposition 5. For this graph, the value $\sigma\,(s, I_G, c)$ follows from Proposition 1—it is equal to $|c_{i_1}| \cdot \sigma_{i_1} + |c_{i_2}| \cdot \sigma_{i_2}$. Thus, by Proposition 3, we get the desired result.

Proof of Propositions 6 is similar to the proof of Proposition 5.

Proof of Proposition 7. Since the vertices i_2 and i_3 are not connected, this means that the measurement errors corresponding to these vertices are independent, so the length of $\mathbf{v}_{i_2} + \mathbf{v}_{i_3}$ is equal to $\sqrt{\|\mathbf{v}_{i_2}\|^2 + \|\mathbf{v}_{i_3}\|^2}$.

The vertex i_1 is connected to both i_2 and i_3, which means that we know nothing about the dependence between the corresponding measurement errors. Thus, as we have described earlier, the largest possible length of the sum

$$\mathbf{v}_{i_1} + \mathbf{v}_{i_2} + \mathbf{v}_{i_3} = \mathbf{v}_{i_1} + \left(\mathbf{v}_{i_2} + \mathbf{v}_{i_3}\right)$$

can be obtained by adding the lengths of $\mathbf{v}_{i_1}$ and of $\mathbf{v}_{i_2} + \mathbf{v}_{i_3}$:

$$\|\mathbf{v}_{i_3}\| + \sqrt{\|\mathbf{v}_{i_2}\|^2 + \|\mathbf{v}_{i_3}\|^2}.$$

The desired result now follows from Proposition 3.

Proof of Proposition 8 is similar to the proof of Proposition 7.

Proof of Proposition 9. We have no restriction on vectors $\mathbf{v}_{i_2}$ and $\mathbf{v}_{i_3}$, so the largest possible length of their sum $\mathbf{v}_{i_2} + \mathbf{v}_{i_3}$ is the sum of their lengths: $\|\mathbf{v}_{i_2}\| + \|\mathbf{v}_{i_3}\|$. There is no edge between i_4 and the group of vertices (i_2, i_3); this means that the measurement errors corresponding to i_4 are independent from the measurement errors Δx_{i_2} and Δx_{i_3}. Thus, the vector $\mathbf{v}_{i_4}$ is orthogonal to vectors $\mathbf{v}_{i_2}$ and $\mathbf{v}_{i_3}$ and is, thus, orthogonal to their sum $\mathbf{v}_{i_2} + \mathbf{v}_{i_3}$. So, the maximum length of the sum

$$\mathbf{v}_{i_2} + \mathbf{v}_{i_3} + \mathbf{v}_{i_4} = \left(\mathbf{v}_{i_2} + \mathbf{v}_{i_3}\right) + \mathbf{v}_{i_4}$$

is equal to the square root of the sums of their lengths:

$$\sqrt{\left(\|\mathbf{v}_{i_2}\| + \|\mathbf{v}_{i_3}\|\right)^2 + \|\mathbf{v}_{i_4}\|^2}.$$

Since i_1 is connected to all the three other vertices, this means that there is no restriction on the relation between the vector i_1 and three other vectors. So, the maximum length of the sum

$$\mathbf{v}_{i_1} + \mathbf{v}_{i_2} + \mathbf{v}_{i_3} + \mathbf{v}_{i_4} = \mathbf{v}_{i_1} + \left(\mathbf{v}_{i_2} + \mathbf{v}_{i_3} + \mathbf{v}_{i_4}\right).$$

is equal to the sum of the lengths:

$$\|\mathbf{v}_{i_1}\| + \sqrt{\left(\|\mathbf{v}_{i_2}\| + \|\mathbf{v}_{i_3}\|\right)^2 + \|\mathbf{v}_{i_4}\|^2}.$$

The use of Proposition 3 completes the proof.

Proof of Proposition 10. Vectors i_2 and i_4 are independent, so the length of the sum $\mathbf{v}_{i_2} + \mathbf{v}_{i_4}$ is equal to $\sqrt{\|\mathbf{v}_{i_2}\|^2 + \|\mathbf{v}_{i_4}\|^2}$. Now, there are no restrictions on the relation between $\mathbf{v}_{i_1}$, $\mathbf{v}_{i_3}$, and $\mathbf{v}_{i_2} + \mathbf{v}_{i_4}$. Thus, the maximum length of the sum

$$\mathbf{v}_{i_1} + \mathbf{v}_{i_2} + \mathbf{v}_{i_3} + \mathbf{v}_{i_4} = \mathbf{v}_{i_1} + \mathbf{v}_{i_3} + \left(\mathbf{v}_{i_2} + \mathbf{v}_{i_4}\right)$$

is equal to the sum of the lengths:

$$\|\mathbf{v}_{i_1}\| + \|\mathbf{v}_{i_3}\| + \sqrt{\|\mathbf{v}_{i_2}\|^2 + \|\mathbf{v}_{i_4}\|^2}.$$

The use of Proposition 3 completes the proof.

Proof of Proposition 11 is similar to the proofs of Propositions 5 and 6.

Proof of Proposition 12. The given graph means that between the four vertices, the only independent pairs are those which are not connected by an egde, i.e., pairs (i_3, i_1), (i_1, i_4), and (i_4, i_2). One can easily see that these vertices also form a linear graph. For this case, the largest value of the sum of the four vectors is computed in the proof of Proposition 23. The use of Proposition 3 completes the proof.

Proof of Proposition 13. Since the vertices i_1 and i_3 are not connected, this means that the measurement errors corresponding to these vertices are independent, so the length of $\mathbf{v}_{i_1} + \mathbf{v}_{i_3}$ is equal to $\sqrt{\|\mathbf{v}_{i_1}\|^2 + \|\mathbf{v}_{i_3}\|^2}$. Similarly, since the vertices i_2 and i_4 are not connected, this means that the measurement errors corresponding to these vertices are independent, so the length of $\mathbf{v}_{i_2} + \mathbf{v}_{i_4}$ is equal to $\sqrt{\|\mathbf{v}_{i_2}\|^2 + \|\mathbf{v}_{i_4}\|^2}$.

Each of the vertices i_1 and i_3 is connected to both i_2 and i_4, which means that we know nothing about the dependence between the corresponding measurement errors. Thus, as we have described earlier, the largest possible length of the sum

$$\mathbf{v}_{i_1} + \mathbf{v}_{i_2} + \mathbf{v}_{i_3} + \mathbf{v}_{i_4} = \left(\mathbf{v}_{i_1} + \mathbf{v}_{i_3}\right) + \left(\mathbf{v}_{i_2} + \mathbf{v}_{i_4}\right)$$

can be obtained by adding the lengths of $\mathbf{v}_{i_1} + \mathbf{v}_{i_3}$ and of $\mathbf{v}_{i_2} + \mathbf{v}_{i_4}$:

$$\sqrt{\|\mathbf{v}_{i_1}\|^2 + \|\mathbf{v}_{i_3}\|^2} + \sqrt{\|\mathbf{v}_{i_2}\|^2 + \|\mathbf{v}_{i_4}\|^2}.$$

The desired result now follows from Proposition 3.

Proof of Proposition 14. The proof of this proposition naturally follows from the geometric interpretation, in which we associate a vector to each random variable, and we are looking for a configuration in which the sum of these vectors has the largest length. Here, the sum $\mathbf{v}$ of all the corresponding vectors $\mathbf{v} = \mathbf{v}_1 + \cdots + \mathbf{v}_n$ can be represented as

$$\sum_{i \notin V} \mathbf{v}_i + \mathbf{a},$$

where

$$\mathbf{a} \stackrel{\text{def}}{=} \sum_{j \in V} \mathbf{v}_j.$$

We do not have any restrictions on the relative orientation of the vectors $\mathbf{v}_i$ corresponding to "normal" errors ($i \notin V$) and of the vector $\mathbf{a}$. Thus, the largest possible value of the length of the sum $\mathbf{v}$ is equal to the sum of the lengths of the "normal" vectors $\mathbf{v}_i$ and of the vector $\mathbf{a}$:

$$\max \|\mathbf{v}\| = \sum_{i \notin V} \|\mathbf{v}_i\| + \|\mathbf{a}\|.$$

The values $\|\mathbf{v}_i\|$ are given: they are equal to $|c_i| \cdot \sigma_i$. Thus, the largest possible value of $\|\mathbf{v}\|$ is attained when the length $\|\mathbf{a}\|$ is the largest. This largest length is what in Definitions 1 and 4 we denoted by $\sigma\,(s_V, D_G, c_V)$. Thus, we get the desired formula.

The proposition is proven.

Proof of Proposition 16. For this graph, the value $\sigma\,(s, D_G, c)$ follows from Proposition 2—it is equal to $\sqrt{c_{i_1}^2 \cdot \sigma_{i_1}^2 + c_{i_2}^2 \cdot \sigma_{i_2}^2}$. Thus, by Proposition 12, we get the desired result.

Proof of Proposition 17 is similar to the proof of Proposition 16.

Proof of Proposition 18. Since the vertices i_2 and i_3 are not connected, this means that we do not have any restrictions on the relative location of the vectors $\mathbf{v}_{i_2}$ and $\mathbf{v}_{i_3}$, so the largest possible value of the length of the sum $\mathbf{v}_{i_2} + \mathbf{v}_{i_3}$ is equal to the sum of the lengths $\|\mathbf{v}_{i_2}\| + \|\mathbf{v}_{i_3}\|$.

The vertex i_1 is connected to both i_2 and i_3, which means that the measurement error corresponding to i_1 is independent of the errors corresponding to i_2 and i_3. Thus, as we have described earlier, the largest possible length of the sum

$$\mathbf{v}_{i_1} + \mathbf{v}_{i_2} + \mathbf{v}_{i_3} = \mathbf{v}_{i_1} + \left(\mathbf{v}_{i_2} + \mathbf{v}_{i_3}\right)$$

is equal to

$$\sqrt{\|\mathbf{v}_{i_1}\|^2 + \left(\|\mathbf{v}_{i_2}\| + \|\mathbf{v}_{i_3}\|\right)^2}.$$

The desired result now follows from Proposition 14.

Proof of Proposition 19 is similar to the proof of Proposition 18.

Proof of Proposition 20. The vectors $\mathbf{v}_{i_2}$ and $\mathbf{v}_{i_3}$ are independent, so the length of their sum $\mathbf{v}_{i_2} + \mathbf{v}_{i_3}$ is equal to $\sqrt{\|\mathbf{v}_{i_2}\|^2 + \|\mathbf{v}_{i_3}\|^2}$. There is no edge between i_4 and the group of vertices (i_2, i_3), this means there is no restriction on the relation between $\mathbf{v}_{i_4}$ and $\mathbf{v}_{i_2} + \mathbf{v}_{i_3}$. Thus, the largest possible length of the sum

$$\mathbf{v}_{i_2} + \mathbf{v}_{i_3} + \mathbf{v}_{i_4} = \left(\mathbf{v}_{i_2} + \mathbf{v}_{i_3}\right) + \mathbf{v}_{i_4}$$

is equal to the sum of their lengths:

$$\sqrt{\|\mathbf{v}_{i_2}\|^2 + \|\mathbf{v}_{i_3}\|^2} + \|\mathbf{v}_{i_4}\|.$$

Since i_1 is connected to all the three other vertices, this means that this vector is orthogonal to three other vectors—and thus, to their sum. So, the maximum length of the sum

$$\mathbf{v}_{i_1} + \mathbf{v}_{i_2} + \mathbf{v}_{i_3} + \mathbf{v}_{i_4} = \mathbf{v}_{i_1} + \left(\mathbf{v}_{i_2} + \mathbf{v}_{i_3} + \mathbf{v}_{i_4}\right).$$

is equal to

$$\sqrt{\|\mathbf{v}_{i_1}\|^2 + \left(\sqrt{\|\mathbf{v}_{i_2}\|^2 + \|\mathbf{v}_{i_3}\|^2} + \|\mathbf{v}_{i_4}\|\right)^2}.$$

The use of Proposition 11 completes the proof.

Proof of Proposition 21. There is no constraint on the vectors i_2 and i_4, so the maximum length of the sum $\mathbf{v}_{i_2} + \mathbf{v}_{i_4}$ is equal to the sum of their length: $\|\mathbf{v}_{i_2}\| + \|\mathbf{v}_{i_4}\|$. Now, the three vectors $\mathbf{v}_{i_1}$, $\mathbf{v}_{i_3}$, and $\mathbf{v}_{i_2} + \mathbf{v}_{i_4}$ are independent. Thus, the maximum length of the sum

$$\mathbf{v}_{i_1} + \mathbf{v}_{i_2} + \mathbf{v}_{i_3} + \mathbf{v}_{i_4} = \mathbf{v}_{i_1} + \mathbf{v}_{i_3} + \left(\mathbf{v}_{i_2} + \mathbf{v}_{i_4}\right)$$

is equal to:

$$\sqrt{\|\mathbf{v}_{i_1}\|^2 + \|\mathbf{v}_{i_3}\|^2 + \left(\|\mathbf{v}_{i_2}\| + \|\mathbf{v}_{i_4}\|\right)^2}.$$

The use of Proposition 11 completes the proof.

Proof of Proposition 22 is similar to the proof of Propositions 16 and 17.

Proof of Proposition 23. In accordance with Proposition 14, we need to compute $\sigma \stackrel{\text{def}}{=} \sigma\,(s_V, D_G, c_V)$, the largest possible length of the vector $\mathbf{v} = \mathbf{v}_{i_1} + \mathbf{v}_{i_2} + \mathbf{v}_{i_3} + \mathbf{v}_{i_4}$. In general, the square $\|\mathbf{v}\|^2$ of the length $\|\mathbf{v}\|$ of the sum $\mathbf{v} = \mathbf{v}_{i_1} + \mathbf{v}_{i_2} + \mathbf{v}_{i_3} + \mathbf{v}_{i_4}$ of the four vectors is equal to

$$\|\mathbf{v}\|^2 = \|\mathbf{v}_{i_1}\|^2 + \|\mathbf{v}_{i_2}\|^2 + \|\mathbf{v}_{i_3}\|^2 + \|\mathbf{v}_{i_4}\|^2 +$$

$$2\mathbf{v}_{i_1} \cdot \mathbf{v}_{i_2} + 2\mathbf{v}_{i_1} \cdot \mathbf{v}_{i_3} + 2\mathbf{v}_{i_1} \cdot \mathbf{v}_{i_4} + 2\mathbf{v}_{i_2} \cdot \mathbf{v}_{i_3} + 2\mathbf{v}_{i_2} \cdot \mathbf{v}_{i_4} + 2\mathbf{v}_{i_3} \cdot \mathbf{v}_{i_4}.$$

For the situation described by the given graph, vector $\mathbf{v}_{i_1}$ is orthogonal to $\mathbf{v}_{i_2}$, the vector $\mathbf{v}_{i_2}$ is orthogonal to $\mathbf{v}_{i_3}$, and the vector $\mathbf{v}_{i_3}$ is orthogonal to $\mathbf{v}_{i_4}$. Thus, we have

$$\|\mathbf{v}\|^2 = \|\mathbf{v}_{i_1}\|^2 + \|\mathbf{v}_{i_2}\|^2 + \|\mathbf{v}_{i_3}\|^2 + \|\mathbf{v}_{i_4}\|^2 + 2\mathbf{v}_{i_1} \cdot \mathbf{v}_{i_3} + 2\mathbf{v}_{i_1} \cdot \mathbf{v}_{i_4} + 2\mathbf{v}_{i_2} \cdot \mathbf{v}_{i_4}.$$

The length of each vector $\|\mathbf{v}_{i_j}\|$ is fixed $\|\mathbf{v}_{i_j}\|^2 = v_{i_j}^2$, so to maximize the length of the sum, we need to maximize the sum of the remaining terms:

$$2\mathbf{v}_{i_1} \cdot \mathbf{v}_{i_3} + 2\mathbf{v}_{i_1} \cdot \mathbf{v}_{i_4} + 2\mathbf{v}_{i_2} \cdot \mathbf{v}_{i_4}.$$

Let us denote the half of this sum by J, then the sum itself becomes equal to $2J$.

We need to maximize the sum $2J$ under the constraints

$$\|\mathbf{v}_{i_j}\|^2 = v_{i_j}^2 \text{ for all } j,\ \mathbf{v}_{i_1} \cdot \mathbf{v}_{i_2} = 0,\ \mathbf{v}_{i_2} \cdot \mathbf{v}_{i_3} = 0, \text{ and } \mathbf{v}_{i_3} \cdot \mathbf{v}_{i_4} = 0.$$

By using the Lagrange multiplier method, we can reduce the above-described conditional optimization problem to the following unconstrained optimization problem:

$$2\mathbf{v}_{i_1}\cdot\mathbf{v}_{i_3}+2\mathbf{v}_{i_1}\cdot\mathbf{v}_{i_4}+2\mathbf{v}_{i_2}\cdot\mathbf{v}_{i_4}+\sum_{j=1}^{4}\lambda_j\cdot\|\mathbf{v}_{i_j}\|^2+\sum_{j=1}^{3}\mu_j\cdot\mathbf{v}_{i_j}\cdot\mathbf{v}_{i_{j+1}},$$

where λ_j and μ_j are Lagrange multipliers.

Differentiating this expression with respect to $\mathbf{v}_{i_2}$ and equating the derivative to 0, we conclude that

$$2\mathbf{v}_{i_4}+2\lambda_2\cdot\mathbf{v}_{i_2}+\mu_1\cdot\mathbf{v}_{i_1}+\mu_2\cdot\mathbf{v}_{i_3}=0,$$

hence

$$\mathbf{v}_{i_4}=-\lambda_2\cdot\mathbf{v}_{i_2}-\frac{1}{2}\cdot\mu_1\cdot\mathbf{v}_{i_1}-\frac{1}{2}\cdot\mu_2\cdot\mathbf{v}_{i_3}.$$

So, the vector $\mathbf{v}_{i_4}$ belongs to the linear space generated by vectors $\mathbf{v}_{i_2}$, $\mathbf{v}_{i_3}$, and $\mathbf{v}_{i_1}$. Let us denote the unit vectors in the directions of $\mathbf{v}_{i_2}$ and $\mathbf{v}_{i_3}$ by, correspondingly,

$$\mathbf{e}_2=\frac{\mathbf{v}_{i_2}}{v_{i_2}}\text{ and }\mathbf{e}_3=\frac{\mathbf{v}_{i_3}}{v_{i_3}}.$$

Since the vectors $\mathbf{v}_{i_2}$ and $\mathbf{v}_{i_3}$ are orthogonal, the unit vectors $\mathbf{e}_2$ and $\mathbf{e}_3$ are orthogonal too, so they can be viewed as two vectors from the orthonormal basis in the linear space generated by the vectors $\mathbf{v}_{i_2}$, $\mathbf{v}_{i_3}$, and $\mathbf{v}_{i_1}$.

- If this linear space is 3-dimensional, in this 3-D space we can select the third unit vector $\mathbf{e}$ which is orthogonal to both $\mathbf{e}_2$ and $\mathbf{e}_3$.
- If the above linear space is 2-dimensional—i.e., if $\mathbf{v}_{i_1}$ lies in the 2-D space generated by $\mathbf{v}_{i_2}$ and $\mathbf{v}_{i_3}$—then let us take, as $\mathbf{e}$, any unit vector which is orthogonal to both $\mathbf{e}_2$ and $\mathbf{e}_3$.

In both cases, vectors $\mathbf{v}_{i_1}$ and $\mathbf{v}_{i_4}$ belong to the linear space generated by the vectors $\mathbf{e}_2$, $\mathbf{e}_3$, and $\mathbf{e}$.

In particular, this means that $\mathbf{v}_{i_1}=c_{12}\cdot\mathbf{e}_2+c_{13}\cdot\mathbf{e}_3+c_1\cdot\mathbf{e}$ for some numbers c_{12}, c_{13}, and c_1. Since $\mathbf{v}_{i_1}\perp\mathbf{v}_{i_2}$, we have $c_{12}=0$, so $\mathbf{v}_{i_1}=c_{13}\cdot\mathbf{e}_3+c_1\cdot\mathbf{e}$. From this formula, we conclude that $v_{i_1}^2=\|\mathbf{v}_{i_1}\|^2=c_{13}^2+c_1^2$, so $c_1^2\le v_{i_1}^2$. Let us denote the ratio c_1/v_{i_1} by β_1, then $c_1=v_{i_1}\cdot\beta_1$ and, correspondingly, $c_{13}=v_{i_1}\cdot\sqrt{1-\beta_1^2}$. So, the expression for $\mathbf{v}_{i_1}$ takes the form

$$\mathbf{v}_{i_1}=v_{i_1}\cdot\sqrt{1-\beta_1^2}\cdot\mathbf{e}_3+v_{i_1}\cdot\beta_1\cdot\mathbf{e}.$$

Similarly, we can conclude that

$$\mathbf{v}_{i_4}=v_{i_4}\cdot\sqrt{1-\beta_4^2}\cdot\mathbf{e}_2+v_{i_4}\cdot\beta_4\cdot\mathbf{e},$$

for some value β_4 for which $|\beta_4| \le 1$. For each pair of orthogonal vectors $\mathbf{e}_2$ and $\mathbf{v}_{i_3}$ of lengths v_{i_2} and v_{i_3}, the above-defined vectors satisfy all the constraints. So, what remains is to find the values β_1 and β_4 for which the expression

$$2\mathbf{v}_{i_1} \cdot \mathbf{v}_{i_3} + 2\mathbf{v}_{i_1} \cdot \mathbf{v}_{i_4} + 2\mathbf{v}_{i_2} \cdot \mathbf{v}_{i_4}$$

attaints its largest value—i.e., equivalently, for which the above-defined half-of-the-maximized-expression

$$J = \mathbf{v}_{i_1} \cdot \mathbf{v}_{i_3} + \mathbf{v}_{i_1} \cdot \mathbf{v}_{i_4} + \mathbf{v}_{i_2} \cdot \mathbf{v}_{i_4}$$

attains its largest value. Substituting the above expressions for $\mathbf{v}_{i_1}$ and $\mathbf{v}_{i_4}$ into this formula, and taking into account that, by our choice of $\mathbf{e}_2$ and $\mathbf{e}_3$, we have $\mathbf{v}_{i_2} = v_{i_2} \cdot \mathbf{e}_2$ and $\mathbf{v}_{i_3} = v_{i_3} \cdot \mathbf{e}_3$, we conclude that

$$J = v_{i_1} \cdot v_{i_3} \cdot \sqrt{1 - \beta_1^2} + v_{i_2} \cdot v_{i_4} \cdot \sqrt{1 - \beta_4^2} + \beta_1 \cdot \beta_4 \cdot v_{i_1} \cdot v_{i_4}.$$

Each of the unknown β_1 and β_4 has values from the interval $[-1, 1]$. Thus, for each of the variables β_1 and β_4, the maximum of this expression is attained:

- either at one of the endpoints -1 and 1 of this interval,
- or at the point inside this interval, in which case the derivative with respect to this variable should be equal to 0.

We have two cases for each of the two variables β_1 and β_4, so overall, we need to consider all $2 \cdot 2 = 4$ cases. To find the largest possible value of the expression J, we need to consider all four possible cases, and find the largest of the corresponding values. Let us consider these cases one by one.

Case 1. If both values β_1 and β_4 are equal to ± 1, then we get $J = \pm v_{i_1} \cdot v_{i_4}$. The largest of these values is when the sign is positive, then the value of the quantity J is equal to $J_1 = v_{i_1} \cdot v_{i_4}$.

Case 2. Let us now consider the case when $\beta_1 = \pm 1$ and $\beta_4 \in (-1, 1)$. In this case, the expression J takes the form $J = v_{i_2} \cdot v_{i_4} \cdot \sqrt{1 - \beta_4^2} \pm v_{i_1} \cdot v_{i_4} \cdot \beta_4$. Differentiating this expression with respect to β_4 and equating the derivative to 0, we get

$$-\frac{2\beta_4 \cdot v_{i_2} \cdot v_{i_4}}{2\sqrt{1 - \beta_4^2}} \pm v_{i_1} \cdot v_{i_4} \cdot \beta_1 = 0.$$

If we divide both sides by v_{i_4}, divide both the numerator and the denominator of the fraction by a common factor 2, and multiply both sides by the denominator, we get

$$\beta_4 \cdot v_{i_2} = \pm\sqrt{1 - \beta_4^2} \cdot v_{i_1}.$$

If we square both sides, we get

$$\beta_4^2 \cdot v_{i_2}^2 = \left(1 - \beta_4^2\right) \cdot v_{i_1}^2 = v_{i_1}^2 - \beta_4^2 \cdot v_{i_1}^2.$$

So

$$\beta_4^2 \cdot \left(v_{i_1}^2 + v_{i_2}^2\right) = v_{i_2}^2$$

and

$$\beta_4^2 = \frac{v_{i_2}^2}{v_{i_1}^2 + v_{i_2}^2}.$$

Therefore,

$$1 - \beta_4^2 = \frac{v_{i_1}^2}{v_{i_1}^2 + v_{i_2}^2},$$

so

$$\beta_4 = \pm \frac{v_{i_2}}{\sqrt{v_{i_1}^2 + v_{i_2}^2}}$$

and

$$\sqrt{1 - \beta_4^2} = \pm \frac{v_{i_1}}{\sqrt{v_{i_1}^2 + v_{i_2}^2}}.$$

Substituting these expressions into the formula for J, we conclude that

$$J = \pm \frac{v_{i_2}^2 \cdot v_{i_4}}{\sqrt{v_{i_1}^2 + v_{i_2}^2}} \pm \frac{v_{i_1}^2 \cdot v_{i_4}}{\sqrt{v_{i_1}^2 + v_{i_2}^2}}.$$

The largest value of this expression is attained when both signs are positive, so we get

$$J = \frac{v_{i_2}^2 \cdot v_{i_4}}{\sqrt{v_{i_1}^2 + v_{i_2}^2}} + \frac{v_{i_1}^2 \cdot v_{i_4}}{\sqrt{v_{i_1}^2 + v_{i_2}^2}} = \frac{v_{i_4} \cdot \left(v_{i_1}^2 + v_{i_2}^2\right)}{\sqrt{v_{i_1}^2 + v_{i_2}^2}}$$

and thus, the value J is equal to

$$J_2 = v_{i_4} \cdot \sqrt{v_{i_1}^2 + v_{i_2}^2}.$$

In this case, the largest value of $\|\mathbf{v}\|^2$ is equal to:

$$\sigma_2^2 = v_{i_1}^2 + v_{i_2}^2 + v_{i_3}^2 + v_{i_4}^2 + 2v_{i_4} \cdot \sqrt{v_{i_1}^2 + v_{i_2}^2} =$$

$$v_{i_3}^2 + \left(\left(v_{i_1}^2 + v_{i_2}^2 \right) + v_{i_4}^2 + 2v_{i_4} \cdot \sqrt{v_{i_1}^2 + v_{i_2}^2} \right) =$$

$$v_{i_3}^2 + \left(\sqrt{v_{i_1}^2 + v_{i_2}^2} + v_{i_4} \right)^2 .$$

Comparing Case 1 and Case 2. Since $v_{i_1}^2 + v_{i_2}^2 > v_{i_1}^2$, we have $\sqrt{v_{i_1}^2 + v_{i_2}^2} > v_{i_1}$, hence $J_2 = v_{i_4} \cdot \sqrt{v_{i_1}^2 + v_{i_2}^2} > v_{i_1} \cdot v_{i_4} = J_1$. Thus, when we are looking for the largest value of the expression J, we can safely ignore Case 1, since the values obtained in Case 2 can be larger than anything we get in Case 1.

Case 3. Similarly, we can consider the case when $\beta_4 = \pm 1$ and $\beta_1 \in (-1, 1)$. In this case, we get the largest possible value of J equal to $J_3 = v_{i_1} \cdot \sqrt{v_{i_3}^2 + v_{i_4}^2}$, so the largest possible value of σ^2 is equal to:

$$\sigma_3^2 = v_{i_2}^2 + \left(v_{i_1} + \sqrt{v_{i_3}^2 + v_{i_4}^2} \right) .$$

Case 4. Finally, let us consider the case when for the pair (β_1, β_4) at which the expression J attains its largest value, both values β_1 and β_4 are located inside the interval $(-1, 1)$. In this case, to find the maximum of the expression J, we need to differentiate it with respect to the unknowns β_1 and β_4 and equate the resulting derivatives to 0. If we differentiate by β_1, we get

$$-\frac{2\beta_1 \cdot v_{i_1} \cdot v_{i_3}}{2\sqrt{1 - \beta_1^2}} + v_{i_1} \cdot v_{i_4} \cdot \beta_4 = 0.$$

Thus,

$$\beta_4 = \frac{\beta_1 \cdot v_{i_3}}{\sqrt{1 - \beta_1^2} \cdot v_{i_4}},$$

and

$$\beta_4^2 = \frac{\beta_1^2 \cdot v_{i_3}^2}{(1 - \beta_1^2) \cdot v_{i_4}^2}.$$

Differentiating the above expression for J with respect to β_4 and equating the derivative to 0, we conclude that

$$-\frac{2\beta_4 \cdot v_{i_2} \cdot v_{i_4}}{2\sqrt{1-\beta_4^2}} + v_{i_1} \cdot v_{i_4} \cdot \beta_1 = 0.$$

If we divide both sides by v_{i_4}, divide both the numerator and the denominator of the fraction by a common factor 2, and multiply both sides by the denominator, we get

$$\beta_4 \cdot v_{i_2} = \sqrt{1-\beta_4^2} \cdot v_{i_1} \cdot \beta_1.$$

If we square both sides, we get

$$\beta_4^2 \cdot v_{i_2}^2 = \left(1-\beta_4^2\right) \cdot v_{i_1}^2 \cdot \beta_1^2 = v_{i_1}^2 \cdot \beta_1^2 - \beta_4^2 \cdot v_{i_1}^2 \cdot \beta_1^2.$$

Substituting the above expression for β_4^2 into this formula, we get

$$\frac{\beta_1^2 \cdot v_{i_2}^2 \cdot v_2^3}{\sqrt{1-\beta_1^2 \cdot v_{i_4}^2}} = \beta_1^2 \cdot v_{i_1}^2 - \frac{\beta_1^4 \cdot v_{i_1}^2 \cdot v_{i_3}^2}{\sqrt{1-\beta_1^2 \cdot v_{i_4}^2}}.$$

Case 4, subcase when $\beta_1 = 0$. Both sides of this equality contain the common factor β_1. So, it is possible that $\beta_1 = 0$, in which case $\beta_4 = 0$, and the expression J attains the value

$$J_0 = v_{i_1} \cdot v_{i_3} + v_{i_2} \cdot v_{i_4}.$$

In this case, the value of σ^2 is equal to:

$$\sigma_0^2 = v_{i_1}^2 + v_{i_2}^2 + v_{i_3}^2 + v_{i_4}^2 + 2v_{i_1} \cdot v_{i_3} + 2v_{i_2} \cdot v_{i_4} =$$

$$\left(v_{i_1}^2 + v_{i_3}^2 + 2v_{i_1} \cdot v_{i_3}\right) + \left(v_{i_2}^2 + v_{i_4}^2 + 2v_{i_2} \cdot v_{i_4}\right) =$$

$$\left(v_{i_1} + v_{i_3}\right)^2 + \left(v_{i_2} + v_{i_4}\right)^2.$$

Case 4, subcase when $\beta_1 \neq 0$. If $\beta_1 \neq 0$, then we can divide both sides of the above equality by β_1^2. Multiplying both sides by the denominator, we get

$$v_{i_2}^2 \cdot v_{i_3}^2 = v_{i_1}^2 \cdot v_{i_4}^2 \cdot \left(1-\beta_1^2\right) - \beta_1^2 \cdot v_{i_1}^2 \cdot v_{i_3}^2,$$

so

$$v_{i_2}^2 \cdot v_{i_3}^2 = v_{i_1}^2 \cdot v_{i_4}^2 - \beta_1^2 \cdot v_{i_1}^2 \cdot v_{i_4}^2 - \beta_1^2 \cdot v_{i_1}^2 \cdot v_{i_3}^2.$$

If we move all the terms containing β_1^2 to the left-hand side and all the other terms to the right-hand side, we get:

$$\beta_1^2 \cdot v_{i_1}^2 \cdot \left(v_{i_3}^2 + v_{i_4}^2\right) = v_{i_1}^2 \cdot v_{i_4}^2 - v_{i_1}^2 \cdot v_{i_3}^2,$$

thus

$$\beta_1^2 = \frac{v_{i_1}^2 \cdot v_{i_4}^2 - v_{i_2}^2 \cdot v_{i_3}^2}{v_{i_1}^2 \cdot (v_{i_3}^2 + v_{i_4}^2)}.$$

Here:

- when $v_{i_1} \cdot v_{i_4} < v_{i_2} \cdot v_{i_3}$, the right-hand side is negative, so we cannot have such a case;
- when $v_{i_1} \cdot v_{i_4} = v_{i_2} \cdot v_{i_3}$, then $\beta_1 = 0$, and we have already analyzed this case.

So, the only possibility to have $\beta_1 \neq 0$ is when $v_{i_1} \cdot v_{i_4} > v_{i_2} \cdot v_{i_3}$.

In general, the situation does not change if we swap 1 and 4 and swap 2 and 3. Thus, for β_4^2, we get a similar expression

$$\beta_4^2 = \frac{v_{i_1}^2 \cdot v_{i_4}^2 - v_{i_2}^2 \cdot v_{i_3}^2}{v_{i_4}^2 \cdot (v_{i_1}^2 + v_{i_2}^2)}.$$

From the expressions for β_1^2 and β_4^2, we conclude that

$$1 - \beta_1^2 = 1 - \frac{v_{i_1}^2 \cdot v_{i_4}^2 - v_{i_2}^2 \cdot v_{i_3}^2}{v_{i_1}^2 \cdot (v_{i_3}^2 + v_{i_4}^2)} =$$

$$\frac{v_{i_1}^2 \cdot v_{i_3}^2 + v_{i_1}^2 \cdot v_{i_4}^2 - v_{i_1}^2 \cdot v_{i_4}^2 + v_{i_2}^2 \cdot v_{i_3}^2}{v_{i_1}^2 \cdot (v_{i_3}^2 + v_{i_4}^2)} =$$

$$\frac{v_{i_3}^2 \cdot (v_{i_1}^2 + v_{i_2}^2)}{v_{i_1}^2 \cdot (v_{i_3}^2 + v_{i_4}^2)}.$$

Similarly, we have

$$1 - \beta_4^2 = \frac{v_{i_2}^2 \cdot (v_{i_3}^2 + v_{i_4}^2)}{v_{i_4}^2 \cdot (v_{i_1}^2 + v_{i_2}^2)}.$$

Thus, for the expression J, we get the value

$$J_4 = v_{i_1} \cdot v_{i_3} \cdot \frac{v_{i_3} \cdot \sqrt{v_{i_1}^2 + v_{i_2}^2}}{v_{i_1} \cdot \sqrt{v_{i_3}^2 + v_{i_4}^2}} + v_{i_2} \cdot v_{i_4} \cdot \frac{v_{i_2} \cdot \sqrt{v_{i_3}^2 + v_{i_4}^2}}{v_{i_4} \cdot \sqrt{v_{i_1}^2 + v_{i_2}^2}} +$$

$$v_{i_1} \cdot v_{i_4} \cdot \frac{v_{i_1}^2 \cdot v_{i_4}^2 - v_{i_2}^2 \cdot v_{i_3}^2}{v_{i_1} \cdot v_{i_4} \cdot \sqrt{v_{i_3}^2 + v_{i_4}^2} \cdot \sqrt{v_{i_1}^2 + v_{i_2}^2}}.$$

We can somewhat simplify this expression if:

- in the first term, we delete v_{i_1} in the numerator and in the denominator,
- in the second term, we delete v_{i_4} from the numerator and from the denominator, and
- in the third term, we delete both v_{i_1} and v_{i_4} from the numerator and from the denominator.

Then, we get:

$$J_4 = v_{i_3} \cdot \frac{v_{i_3} \cdot \sqrt{v_{i_1}^2 + v_{i_2}^2}}{\sqrt{v_{i_3}^2 + v_{i_4}^2}} + v_{i_2} \cdot \frac{v_{i_2} \cdot \sqrt{v_{i_3}^2 + v_{i_4}^2}}{\sqrt{v_{i_1}^2 + v_{i_2}^2}} + \frac{v_{i_1}^2 \cdot v_{i_4}^2 - v_{i_2}^2 \cdot v_{i_3}^2}{\sqrt{v_{i_3}^2 + v_{i_4}^2} \cdot \sqrt{v_{i_1}^2 + v_{i_2}^2}}.$$

If we bring all the terms to the common denominator $\sqrt{v_{i_3}^2 + v_{i_4}^2} \cdot \sqrt{v_{i_1}^2 + v_{i_2}^2}$, then we get

$$J_4 = \frac{v_{i_3}^2 \cdot \left(v_{i_1}^2 + v_{i_2}^2\right) + v_{i_2}^2 \cdot \left(v_{i_3}^2 + v_{i_4}^2\right) + v_{i_1}^2 \cdot v_{i_4}^2 - v_{i_2}^2 \cdot v_{i_3}^2}{\sqrt{v_{i_3}^2 + v_{i_4}^2} \cdot \sqrt{v_{i_1}^2 + v_{i_2}^2}}.$$

The numerator of this expression has the form

$$v_{i_1}^2 \cdot v_{i_3}^2 + v_{i_2}^2 \cdot v_{i_3}^2 + v_{i_2}^2 \cdot v_{i_3}^2 + v_{i_2}^2 \cdot v_{i_4}^2 + v_{i_1}^2 \cdot v_{i_4}^2 - v_{i_2}^2 \cdot v_{i_3}^2 =$$

$$v_{i_1}^2 \cdot v_{i_3}^2 + v_{i_2}^2 \cdot v_{i_3}^2 + v_{i_2}^2 \cdot v_{i_4}^2 + v_{i_1}^2 \cdot v_{i_4}^2 =$$

$$\left(v_{i_1}^2 + v_{i_2}^2\right) \cdot \left(v_{i_3}^2 + v_{i_4}^2\right).$$

Thus, we get

$$J_4 = \frac{\left(v_{i_1}^2 + v_{i_2}^2\right) \cdot \left(v_{i_3}^2 + v_{i_4}^2\right)}{\sqrt{v_{i_3}^2 + v_{i_4}^2} \cdot \sqrt{v_{i_1}^2 + v_{i_2}^2}},$$

i.e.,

$$J_4 = \sqrt{v_{i_1}^2 + v_{i_2}^2} \cdot \sqrt{v_{i_3}^2 + v_{i_4}^2}.$$

Comparing J_4 with J_2 and J_3. One can easily see that we always have $J_2^2 \le J_4^2$ and $J_3^2 \le J_4^2$, thus $J_2 \le J_4$ and $J_3 \le J_4$. Thus, if the estimate J_4 is possible, there is no need to consider J_2 and J_3, we only need to cosider J_4 and J_0.

Comparing J_4 and J_0. Let us show that we always have $J_0 \le J_4$, i.e.,

$$v_{i_1} \cdot v_{i_3} + v_{i_2} \cdot v_{i_4} \le \sqrt{v_{i_1}^2 + v_{i_2}^2} \cdot \sqrt{v_{i_3}^2 + v_{i_4}^2}.$$

Indeed, this inequality between positive numbers is equivalent to a similar inequality between their squares:

$$v_{i_1}^2 \cdot v_{i_3}^2 + v_{i_2}^2 \cdot v_{i_4}^2 + 2v_{i_1} \cdot v_{i_2} \cdot v_{i_3} \cdot v_{i_4} \le \left(v_{i_1}^2 + v_{i_2}^2\right) \cdot \left(v_{i_3}^2 + v_{i_4}^2\right),$$

i.e.,

$$v_{i_1}^2 \cdot v_{i_3}^2 + v_{i_2}^2 \cdot v_{i_4}^2 + 2v_{i_1} \cdot v_{i_2} \cdot v_{i_3} \cdot v_{i_4} \le v_{i_1}^2 \cdot v_{i_3}^2 + v_{i_1}^2 \cdot v_{i_4}^2 + v_{i_2}^2 \cdot v_{i_3}^2 + v_{i_2}^2 \cdot v_{i_4}^2.$$

Subtracting $v_{i_1}^2 \cdot v_{i_3}^2 + v_{i_2}^2 \cdot v_{i_4}^2$ from both sides of this inequality, we get an equivalent inequality

$$2v_{i_1} \cdot v_{i_2} \cdot v_{i_3} \cdot v_{i_4} \le v_{i_1}^2 \cdot v_{i_4}^2 + v_{i_2}^2 \cdot v_{i_3}^2,$$

i.e., equivalently,

$$0 \le v_{i_1}^2 \cdot v_{i_4}^2 + v_{i_2}^2 \cdot v_{i_3}^2 - 2v_{i_1} \cdot v_{i_2} \cdot v_{i_3} \cdot v_{i_4} = \left(v_{i_1} \cdot v_{i_4} - v_{i_2} \cdot v_{i_3}\right)^2,$$

which is, of course, always true. Thus, when the estimate J_4 is possible, we do not need to consider the value J_0 either: it is sufficient to take $J = J_4$.

Value of σ in case J_4 is possible: conclusion. So, if the value J_4 is possible, we get

$$\sigma^2 = v_{i_1}^2 + v_{i_2}^2 + v_{i_3}^2 + v_{i_4}^2 + 2\sqrt{v_{i_1}^2 + v_{i_2}^2} \cdot \sqrt{v_{i_3}^2 + v_{i_4}^2} =$$

$$\left(v_{i_1}^2 + v_{i_2}^2\right) + \left(v_{i_3}^2 + v_{i_4}^2\right) + 2\sqrt{v_{i_1}^2 + v_{i_2}^2} \cdot \sqrt{v_{i_3}^2 + v_{i_4}^2} =$$

$$\left(\sqrt{v_{i_1}^2 + v_{i_2}^2} + \sqrt{v_{i_3}^2 + v_{i_4}^2}\right)^2,$$

so $\sigma = \sqrt{v_{i_1}^2 + v_{i_2}^2} + \sqrt{v_{i_3}^2 + v_{i_4}^2}$.

General comment. The desired result for this case now follows from Proposition 14.

Proof of Proposition 24. Since the vertices i_2 and i_4 are not connected, this means that we do not have any restrictions on the relative location of the vectors $\mathbf{v}_{i_2}$ and $\mathbf{v}_{i_4}$, so the largest possible value of the length of the sum $\mathbf{v}_{i_2} + \mathbf{v}_{i_4}$ is equal to the sum of the lengths $\|\mathbf{v}_{i_2}\| + \|\mathbf{v}_{i_4}\|$. Similarly, since the vertices i_1 and i_3 are not connected, this means that we do not have any restrictions on the relative location of the vectors $\mathbf{v}_{i_1}$ and $\mathbf{v}_{i_3}$, so the largest possible value of the length of the sum $\mathbf{v}_{i_1} + \mathbf{v}_{i_3}$ is equal to the sum of the lengths $\|\mathbf{v}_{i_1}\| + \|\mathbf{v}_{i_3}\|$.

Each of the vertices i_1 and i_3 is connected to both i_2 and i_4, which means that the measurement errors corresponding to i_1 and i_3 are independent of the errors corresponding to i_2 and i_4. Thus, as we have described earlier, the largest possible length of the sum

$$\mathbf{v}_{i_1} + \mathbf{v}_{i_2} + \mathbf{v}_{i_3} + \mathbf{v}_{i_4} = \left(\mathbf{v}_{i_1} + \mathbf{v}_{i_3}\right) + \left(\mathbf{v}_{i_2} + \mathbf{v}_{i_4}\right)$$

is equal to

$$\sqrt{\left(\|\mathbf{v}_{i_1}\| + \|\mathbf{v}_{i_3}\|\right)^2 + \left(\|\mathbf{v}_{i_2}\| + \|\mathbf{v}_{i_3}\|\right)^2}.$$

The desired result now follows from Proposition 14.

Acknowledgements This work was supported in part by the National Science Foundation grants 1623190 (A Model of Change for Preparing a New Generation for Professional Practice in Computer Science), and HRD-1834620 and HRD-2034030 (CAHSI Includes), and by the AT&T Fellowship in Information Technology.
It was also supported by the program of the development of the Scientific-Educational Mathematical Center of Volga Federal District No. 075-02-2020-1478, and by a grant from the Hungarian National Research, Development and Innovation Office (NRDI).

References

1. L. Jaulin, M. Kiefer, O. Didrit, E. Walter, *Applied Interval Analysis, with Examples in Parameter and State Estimation, Robust Control, and Robotics* (Springer, London, 2001)
2. B.J. Kubica, *Interval Methods for Solving Nonlinear Contraint Satisfaction, Optimization, and Similar Problems: from Inequalities Systems to Game Solutions* (Springer, Cham, Switzerland, 2019)
3. G. Mayer, *Interval Analysis and Automatic Result Verification* (de Gruyter, Berlin, 2017)
4. R.E. Moore, R.B. Kearfott, M.J. Cloud, *Introduction to Interval Analysis* (SIAM, Philadelphia, 2009)
5. S.G. Rabinovich, *Measurement Errors and Uncertainty: Theory and Practice* (Springer, New York, 2005)
6. D.J. Sheskin, *Handbook of Parametric and Nonparametric Statistical Procedures* (Chapman and Hall/CRC, Boca Raton, FL, 2011)

How to Combine Expert Estimates? How to Estimate Probability in the Intersection of Two Populations?

Miroslav Svítek, Olga Kosheleva, Vladik Kreinovich, and Nguyen Hoang Phuong

Abstract In this paper, we consider two different practical problems that turned to be mathematically similar: (1) how to combine two expert-provided probabilities of some event and (2) how to estimate the frequency of a certain phenomenon (e.g., illness) in an intersection of two populations if we know the frequencies in each of these populations. In both cases, we use the maximum entropy approach to come up with a solution.

1 Formulation of the First Problem

The problem. Suppose that the experts E_1 and E_2 provided two estimates p_1 and p_2 for the probability of some event E. We would like to provide a single estimate that takes both estimates into account.

What information we can use. To properly combine the two estimates, it is important to take into account how related are the opinions of the two experts.

- If in all previous situations and in this situation, the experts gave almost identical opinions, this probably means that they use the same technique to provide their estimates. In this case, the opinion of the second expert does not add anything

M. Svítek
Faculty of Transportation Sciences, Czech Technical University in Prague, Konviktska 20, 110 00 Prague 1, Czech Republic
e-mail: svitek@fd.cvut.cz

O. Kosheleva · V. Kreinovich (✉)
University of Texas at El Paso, 500 W. University, El Paso, TX 79968, USA
e-mail: vladik@utep.edu

O. Kosheleva
e-mail: olgak@utep.edu

N. H. Phuong
Artificial Intelligence Division, Information Technology Faculty, Thang Long University, Nghiem Xuan Yem Road, Hoang Mai District, Hanoi, Vietnam

N. H. Phuong and V. Kreinovich (eds.), *Deep Learning and Other Soft Computing Techniques*, Studies in Computational Intelligence 1097,
https://doi.org/10.1007/978-3-031-29447-1_24

new to the opinion of the first expert. So, the combined probability will still be the same value p_1.

- If in the previous situations, the experts' opinions were independent, this means that they use different data and different techniques to estimate the probability. In this case, e.g., if both experts believe that this event is possible, then taking both opinions into account should increase this probability.
- On the other hand, if the expert opinions are negatively correlated, then we do not know whom to believe, and we should not take either of the probabilities seriously. In this case, the combined probability should be close to the do-not-know 0.5 value.

What we do in this paper. In this paper, we show how to come up with a reasonable numerical value of the combined probability.

2 Formulation of the Second Problem

Formulation of the problem. Suppose that we know the frequencies of a certain phenomenon in two different populations: e.g., the frequency of a certain disease in a 50–60 age group and the frequency of this disease in women. Based on these two frequencies, what is the reasonable estimate for the frequency of this disease in the intersection of these two populations—i.e., among women in the 50–60 age bracket; see, e.g., [1] and references therein.

What we do in this paper. In this paper, we show that this problem is mathematically similar to the first one, and thus, all the methods for solving the first problem can be automatically applied to the second problem as well.

3 Formulation of the Problems in Precise Terms

Notations: case of the first problem. In the first problem, let us make the following notations for the random variables:

- Let E be equal to 1 if the event happens and 0 if it does not.
- Let e_1 be equal to 1 if the first expert is correct in a randomly selected situation, and 0 if the first expert is wrong.
- Similarly, let e_2 be equal to 1 if the second expert is correct in a randomly selected situation, and 0 if the second expert is wrong.

What we know and what we want to estimate. What we know:

- We know the conditional probabilities $p(E \mid e_1) = p_1$ and $p(E \mid e_2) = p_2$.
- Based on the analysis of the previous expert opinions, we can estimate the probability $p(e_1)$—by counting how many times the first expert was right.

- Similarly, based on the analysis of the previous expert opinions, we can estimate the probability $p(e_2)$—by counting how many times the second expert was right.
- We can also estimate the probability $p(e_1 \,\&\, e_2)$ that both experts were right.

Based on this information, we want to estimate the probability $p(E)$.

What if we have only one expert? In this case, we know the conditional probability $p_1 = p(E \mid e_1)$ and the probability $p(e_1)$, and we want to estimate the probability $p(E)$.

Second problem. In the second problem, we only consider folks who belong to one (or both) of the two populations. In this case, we consider the following random variables:

- Let E be equal to 1 if a randomly selected element has the desired phenomenon.
- Let e_1 be 1 if a randomly selected element belongs to the first population.
- Let e_2 be 1 if a randomly selected element belongs to the first population.

For this problem, we know the following information:

- We know the frequencies $p(E \mid e_1) = p_1$ and $p(E \mid e_2) = p_2$ of the phenomenon in each population.
- We know the number of elements n_1 in the first population, the number of elements n_2 in the second population, and the number of elements n_{12} that belong to both populations. In this case, the overall number of elements that belongs to both populations is equal to $n_1 + n_2 - n_{12}$.
- Thus, we can estimate the probabilities of e_1, e_2, and $e_1 \,\&\, e_2$ as follows:

$$p(e_1) = \frac{n_1}{n_1 + n_2 - n_{12}}; \quad p(e_2) = \frac{n_2}{n_1 + n_2 - n_{12}}; \quad p(e_1 \,\&\, e_2) = \frac{n_{12}}{n_1 + n_2 - n_{12}}.$$

Based on this information, we want to find the probability $p(E \mid e_1 \,\&\, e_2)$.

4 How These Problems Can be Solved

Let us use the Maximum Entropy approach. Situations in which we only have partial information about probabilities—and thus, several different probability distributions are consistent with our knowledge—are ubiquitous. In such cases, it makes sense not to pretend that our uncertainty is low—and thus, to select the distribution with the largest possible uncertainty.

A natural measure of uncertainty of a probability distribution is the average number of binary ("yes"-"no") questions that we need to ask to fully determine which statements are true and which are false. It is known that this number is equal to Shannon's entropy $S = -\sum P_i \cdot \log_2(P_i)$, where P_i are the probabilities of different possible situations; see, e.g., [2]. Thus, we need to select the distribution with the largest possible entropy. Such a selection is known as the Maximum Entropy approach.

What this means for the first problem. In the first problem, we have three basic statement E, e_1, and e_2. Each of these statements is either true or false. Thus, we have $2^3 = 8$ possible situations:

$$E \,\&\, e_1 \,\&e_2,\quad E \,\&\, e_1 \,\&\neg e_2,\quad E \,\&\, \neg e_1 \,\&e_2,\quad E \,\&\, \neg e_1 \,\&\neg e_2,$$

$$\neg E \,\&\, e_1 \,\&e_2,\quad \neg E \,\&\, e_1 \,\&\neg e_2,\quad \neg E \,\&\, \neg e_1 \,\&e_2,\quad \neg E \,\&\, \neg e_1 \,\&\neg e_2.$$

Let us use the following notations for their probabilities:

$$p_{111} \stackrel{\text{def}}{=} E \,\&\, e_1 \,\&e_2,\quad p_{110} \stackrel{\text{def}}{=} E \,\&\, e_1 \,\&\neg e_2,\quad p_{101} \stackrel{\text{def}}{=} E \,\&\, \neg e_1 \,\&e_2,$$

$$p_{100} \stackrel{\text{def}}{=} E \,\&\, \neg e_1 \,\&\neg e_2,\quad p_{011} \stackrel{\text{def}}{=} \neg E \,\&\, e_1 \,\&e_2,\quad p_{010} \stackrel{\text{def}}{=} \neg E \,\&\, e_1 \,\&\neg e_2,$$

$$p_{001} \stackrel{\text{def}}{=} \neg E \,\&\, \neg e_1 \,\&e_2,\quad p_{000} \stackrel{\text{def}}{=} \neg E \,\&\, \neg e_1 \,\&\neg e_2.$$

These eight probabilities must add up to 1:

$$p_{111} + p_{110} + p_{101} + p_{100} + p_{011} + p_{010} + p_{001} + p_{000} = 1. \tag{1}$$

Since we know the probability $p(e_1)$, the fact that we know the value $p_1 = p(E \mid e_1) = p(E \,\&\, e_1)/p(e_1)$ is equivalent to knowing the probability $p(E \,\&\, e_1) = p_1 \cdot p(e_1)$. In terms of the basic probabilities, the probability $p(E \,\&\, e_1)$ has the form

$$p(E \,\&\, e_1) = p(E \,\&\, e_1 \,\&\, e_2) + p(E \,\&\, e_1 \,\&\, \neg e_2) = p_{111} + p_{110}.$$

Thus, we have

$$p_{111} + p_{110} = p_1 \cdot p(e_1). \tag{2}$$

Similarly, since we know the probability $p(e_2)$, the fact that we know the value $p_2 = p(E \mid e_2) = p(E \,\&\, e_2)/p(e_2)$ is equivalent to knowing the probability $p(E \,\&\, e_2) = p_2 \cdot p(e_2)$. In terms of the basic probabilities, the probability $p(E \,\&\, e_2)$ has the form

$$p(E \,\&\, e_2) = p(E \,\&\, e_1 \,\&\, e_2) + p(E \,\&\, \neg e_1 \,\&\, e_2) = p_{111} + p_{101}.$$

Thus, we have

$$p_{111} + p_{101} = p_2 \cdot p(e_2). \tag{3}$$

Information about the values $p(e_1)$, $p(e_2)$, and $p(e_1 \,\&\, e_2)$ takes the following form:

$$p_{111} + p_{110} + p_{011} + p_{010} = p(e_1); \tag{4}$$

$$p_{111} + p_{101} + p_{011} + p_{001} = p(e_2); \tag{5}$$

$$p_{111} + p_{011} = p(e_1 \,\&\, e_2). \tag{6}$$

So, to find the values p_{ikj}, we need to maximize the entropy

$$S = -p_{111} \cdot \log_2(p_{111}) - p_{110} \cdot \log_2(p_{110}) - p_{101} \cdot \log_2(p_{101}) -$$

$$p_{100} \cdot \log_2(p_{100}) - p_{011} \cdot \log_2(p_{011}) - p_{010} \cdot \log_2(p_{010}) -$$

$$p_{001} \cdot \log_2(p_{001}) - p_{000} \cdot \log_2(p_{000}) \tag{7}$$

under the constraints (1)–(6).

Entropy (6) is a convex function of the probabilities, and the constraints are linear in terms of these probabilities. Thus, we can use the feasible convex optimization algorithms to find the desired probabilities; see, e.g., [2, 3]. Once we find all the probabilities p_{ijk}, we can compute the desired probability $p(E)$ as

$$p(E) = p_{111} + p_{110} + p_{101} + p_{100}. \tag{8}$$

Comments. We can similarly consider the case when we have more than two experts and the case when we only have one expert. In the situation when we have only one expert, we have four possible situations

$$E \,\&\, e_1, \quad E \,\&\, \neg e_1, \quad \neg E \,\&\, e_1, \quad \neg E \,\&\, \neg e_1.$$

Let us denote the probabilities of these situations by

$$p_{11} = p(E \,\&\, e_1), \quad p_{10} = p(E \,\&\, \neg e_1), \quad p_{01} = p(\neg E \,\&\, e_1), \quad p_{00} = p(\neg E \,\&\, \neg e_1).$$

These four probabilities must add up to 1:

$$p_{11} + p_{10} + p_{01} + p_{00} = 1. \tag{9}$$

The available information—i.e., the values $p(e_1)$ and p_1—lead to the following constraints:

$$p_{11} + p_{01} = p(e_1) \tag{10}$$

and

$$p_{11} = p_1 \cdot p(e_1). \tag{11}$$

In this case, we know the value p_{11}—from the equation (11)—and we can, thus, determine the probability p_{01} from the formula (10), as

$$p_{01} = p(e_1) - p_1 \cdot p(e_1) = p(e_1) \cdot (1 - p_1).$$

The only constraint on the remaining two values p_{00} and p_{10} (coming from the condition (9)) is that

$$p_{00} + p_{10} = 1 - (p_{01} + p_{11}) = 1 - p(e_1).$$

In this case, the maximum entropy approach leads to equal values of these two probabilities:

$$p_{00} = p_{10} = \frac{1 - p(e_1)}{2}.$$

Thus, the resulting estimate for the desired probability $p(E) = p_{11} + p_{10}$ has the form

$$p(E) = p_1 \cdot p(e_1) + \frac{1 - p(e_1)}{2} = \frac{1}{2} + p(e_1) \cdot \left(p_1 - \frac{1}{2}\right). \tag{12}$$

This formula can be alternatively reformulated as

$$p(E) - \frac{1}{2} = p(e_1) \cdot \left(p_1 - \frac{1}{2}\right). \tag{13}$$

In other words, we should not take the expert estimate p_1 at face value, we should adjust this estimate based on the expert's track record.

What this means for the second problem. From the mathematical viewpoint, the two problems has similar inputs, the only two differences are as follows:

- first, we only consider objects that belong to one of the populations, so

$$p_{000} = p_{100} = 0; \tag{14}$$

- second, what we want to estimate is different: instead of the probability $p(E)$ (as in the the first problem) we want to estimate the conditional probability $p(E \mid e_1 \,\&\, e_2)$.

Thus, to solve the second problem, we perform the same optimization as in the first problem—with the additional constraint (14)—to find the probabilities p_{ijk}, and then estimate:

$$p(E \mid e_1 \,\&\, e_2) = \frac{p(E \,\&\, e_1 \,\&\, e_2)}{p(e_1 \,\&\, e_2)} = \frac{p_{111}}{p_{011} + p_{111}}. \tag{15}$$

Acknowledgements This work was supported in part by the National Science Foundation grants 1623190 (A Model of Change for Preparing a New Generation for Professional Practice in Computer Science), and HRD-1834620 and HRD-2034030 (CAHSI Includes), and by the AT&T Fellowship in Information Technology.

It was also supported by the program of the development of the Scientific-Educational Mathematical Center of Volga Federal District No. 075-02-2020-1478, and by a grant from the Hungarian National Research, Development and Innovation Office (NRDI).

References

1. P. Baldi, What's hot in uncertain reasoning. The Reason. **16**(3), 23–24 (2022)
2. E.T. Jaynes, G.L. Bretthorst, *Probability Theory: The Logic of Science* (Cambridge University Press, Cambridge, UK, 2003)
3. R.T. Rockafeller, *Convex Analysis* (Princeton University Press, Princeton, NJ, 1997)

The manufacturer's authorised representative in the EU is Springer Nature Customer Service Centre GmbH, Europaplatz 3, 69115 Heidelberg, Germany. If you have any concerns regarding our products, please contact ProductSafety@springernature.com

Printed and bound by CPI Group (UK) Ltd, Croydon, CR0 4YY
28/09/2026
02224077-0006